The Adventures of

AL JONIEC

Ford Drag Racer

**A Story of
MUSTANGS, COBRA JETS,
BATCARS, AIRPLANES, DRAGONS,
HAIRY ONES and 2,000 WORMS**

CHARLES R. MORRIS
WITH AL JONIEC

STANCE & SPEED

ACKNOWLEDGMENTS

MANY THANKS TO Lloyd Wolfe for sharing never-before-seen photos of Al's cars, and Carl Mentz for his knowledge of Cobras and the Spinelli family for the Boss Mustang material.

Published by:

Stance & Speed
St. Paul, MN
www.stanceandspeed.com

Version 1.0 Published April 2018
Version 1.1 Published August 2018

Morris, Charles R.
The Adventures of Al Joniec, Ford Drag Racer

ISBN-13: 978-0-9891149-9-8

Layout and Design by Tom Heffron

CONTENTS

PREFACE

THE STORY YOU ARE ABOUT TO READ IS TRUE. It is based upon the personal recollections and official records that chronicle the life and racing career of Alexander "Al" Joniec, a man who parlayed a youthful interest in the American automobile, his mechanical aptitude, and driving skill to national prominence as a factory-sponsored drag racer. Among his many talents, Al Joniec can tell a story like no other. But, as with many stories from the heyday of racing, some are "not suitable for younger viewers," and best told by the man himself. In this book the reader will find that the majority of Al's story is told in his own words, but just like the old *Dragnet* television series, sometimes "names have been changed to protect the innocent." This book provides a glimpse into the true spirit of hot rodding and what has become known as the golden age of drag racing through the eyes of one of its pioneers, a man who remains a student of the internal combustion engine to this day.

INTRODUCTION

AS A LIFELONG AUTO ENTHUSIAST, drag racer, and race fan I find myself in a unique position at this point in my life. At an early age I had arrived at the conclusion that I was a Ford man and thus became totally immersed in the automotive products built in Dearborn, Michigan. Of particular interest were the performance accomplishments of my favorite car company. Much like so many young men of my generation, my sports heroes were those individuals who drove race cars (Ford products) in various forms of competition.

Reading about and attending stock car and Indy car races provided me with a glimpse into a world in which my circumstances would never allow me to participate, but drag racing was a different matter altogether. Well before reaching legal driving age, I was going to drag races and assisting older friends when I could. Along the way I became aware of a successful Ford racer from the Philadelphia area named Al Joniec. Unbeknownst to Al at the time, I was one of many teenaged Ford fans cheering for him from the grandstands as he faced off against the drivers of other factory-supported drag cars in the match races that were heavily advertised and wildly popular in the mid-1960s.

Unlike today, where the professional drag race teams are only viewed by spectators from behind barriers in the pits, fans of my era could get up close and personal with their favorite drivers and crews as they worked on cars between rounds of competition. On several occasions I found myself in close proximity to Al, but I merely observed what he was doing, never thinking to engage him in conversation. By the mid-1970s my beloved sport of drag racing was undergoing changes, including the end of factory support from the auto manufacturers. Many top drivers, Al Joniec among them, chose a different path of endeavor and left drag racing behind. My participation as a driver also ended in the mid-1970s, although I remained involved in the sport as a crew member on an alcohol Funny Car. While I never forgot the great memories Al Joniec provided me as a race fan, I had no personal knowledge of his whereabouts for many years.

I had always felt that many popular automotive publications offered up what I believed to be a thinly disguised bias when reporting the racing accomplishments of my favorite car brand. In the late 1980s I decided to begin writing down my memories of drag racing history in a series of magazine articles. Encouraged with the positive response, I decided to attempt writing a book that would chronicle Ford drag racing during the 1960s. That process began by reaching out to those racers who had made the history that I intended to write about and, of course, Al Joniec was one of them.

Upon contacting Al and explaining my purpose, he was taken aback by the fact that anyone was interested in races run three decades previous, or in the value of preserving the recollections of the participants. Once he got past the initial shock of being contacted by a middle-aged guy who seemed interested not just in hearing his stories, but also writing them down for publication, Al proved himself to be a great resource and a storyteller second to none. That first contact with Al eventually led to an enduring friendship, which put me in the unique position of becoming close friends with one of the racing heroes of my youth.

CHAPTER ONE

THE RIGHT TIME, PLACE AND PEOPLE

AL JONIEC GREW UP in the 1950s, a time when many of America's youth might best be described as having 30-weight motor oil running through their veins. Americans were benefitting from an economic resurgence that made it possible for many households to afford the luxury of an automobile. The prospect of obtaining one's own set of wheels offered a sense of freedom heretofore unknown. Al Joniec has likened this feeling of freedom to what it must have been like to own a horse in the old West.

This love of cars would give rise to what were called—at first derisively—"hot rods," loosely defined as secondhand automobiles that had been modified from stock by as a means of displaying one's individuality. The modifications most often found in the hot rod of the time consisted of "speed equipment"—specialty parts that enhanced the horsepower output of a car's engine. The Ford flathead V-8, the mass-produced marvel first introduced in 1932, was perhaps the most popular engine to receive such treatment, in that it could be readily obtained for very little investment and lent itself so well to performance-enhancing modifications. An entire industry was built around enhancing the flathead, with companies such as Edelbrock, Offenhauser, Iskendarian, Fenton, Navarro, and others spearheading the development of parts to feed the hot rod generation's seemingly insatiable need for speed.

Often overlooked was that fact that by tearing apart, rebuilding, modifying, and racing cars that were often resurrected from a junkyard, countless young men were developing valuable mechanical skills that would in time gain the attention of America's auto manufacturers. These corporations eventually began to develop performance-oriented models to appeal to the ever-growing youth market.

With the realization that race track successes led to showroom sales of their products, Ford, General Motors, Chrysler, and even Hudson began to engineer in-house performance parts. They also began to support the efforts of competitors in organized racing events and even fielded racing teams that included factory engineers, technicians, and professional drivers. A horsepower war was being waged by the Big Three auto manufacturers and it appeared that the sky was the limit.

A secondary and under-appreciated effect of Detroit's entry into the world of competitive automotive trials of speed and endurance was the development of components that made production automobiles run, handle, and brake more efficiently. This paid dividends in highway safety and caused automotive engineers to labor toward the development of even better vehicles.

Most of those who describe themselves as car enthusiasts can easily conjure up a vision of that special moment in time when they became enamored with all things automotive. In the case of Al Joniec it was ride in a friend's 1948 Chevy at age 13 that tripped that switch. Possessing no knowledge of the mechanical aspects of an automobile, young Al still found himself fascinated with the functions of this remarkable machine, particularly how the engine started and how the vacuum-assisted, steering column-mounted shift lever changed gears. When Al was 14, an inebriated uncle put Al behind the wheel of a car for the first time. Being a quick study and having carefully observed the skills required to operate an automobile, the teenager passed his first driving test—albeit an unorthodox one—without causing any damage, though he almost sent both himself and his less-than-alert relative through the windshield due to his unfamiliarity with the car's power assisted brakes. Having been bitten by the car bug, Al took to borrowing his cousin's 1937 Buick and honing his driving skills with forays around the block, until the cousin found out and took away the keys to the car.

By age 15, Al's interest in the functioning of this engineering marvel called the internal combustion engine led him to request an explanation from a grizzled local mechanic. The man's lesson, as direct and simplistic as it was, has remained indelibly burned into Al's memory for more than six decades. The complexities of how an engine works were broken down by the old mechanic as, "Suck, Squeeze, Bang, Blow," to describe the intake, compression, ignition, and exhaust strokes of the engine's internal assemblies.

Young Al worked for his uncle to earn some spending money. His uncle would pay him one cent for each worm he could dig up. After delivering 2,000 of the squirming critters to his uncle over a period of time, he had saved up enough money to purchase a car.

To those of us who grew up during the heyday of the car culture, obtaining one's first automobile was an event as significant as falling in love for the first time, and Al Joniec was no exception. As he was about to reach the legal driving age of 16, the matter of foremost importance to the young man was that first car. With a degree of clarity fully understood and appreciated by this author, Al described how he

Al Joniec's first car was a 1940 Ford coupe, one of Ford's all-time classic designs that is still sought-after by hot rodders and collectors today. Al used the proceeds from the gathering of 2,000 worms and a loan from his grandmother to begin his journey to fame as a racer. (Tom Pado photo)

set out to buy a well-worn 1940 Ford coupe from his friend Bill Simon. The Ford was offered at the bargain price of $25 because it had a broken axle. Al, whose entire fortune at the time consisted of $20 earned digging the aforementioned worms, vividly recalls begging his grandmother, on his knees, for the balance.

Once he paid for his prize, Al was faced with the task of making it roadworthy once again. "I remember lying in around two feet of snow with the car jacked up on milk crates as I took the rear apart," Al says. "The black grease was dripping into the snow and covering my clothes and hair. I can still smell it. I had only monkey or adjustable-type wrenches to work with—a ratchet wrench was only for the pros—and some very primitive, old, worn-out open ends borrowed from my uncle. I went to the junkyard in the dead of winter while it was snowing and had to do the same thing to get the replacement axle. I was thrilled with it all and nothing was going to stop me from getting my car running. When I finally got to drive the beat up old '40 I was just bursting with pride and thrilled to death."

A part-time job in a gas station did little to teach the budding hot rodder about mechanical work, but it did supply modest funds, which were immediately invested in his car. The effort required to earn these dollars provided a lesson in service and a work ethic that would pay dividends in the years to come. The young Philadelphian was so absorbed with the task of making his mount better that he even found it necessary to part ways with his serious girlfriend of the time in order to devote more time and money to the car.

During 1953 and 1954, Al's flathead V-8-powered '40 coupe took him to victory many times at the local street racing spot known as the Meadows, a stretch of Delaware Avenue near the intersection of Castor Avenue paralleling the Delaware river. But in 1955, Chevy introduced its 265 cubic inch overhead-valve V-8 and the writing was on the wall for the flathead Ford's dominance. The purchase of a flathead V-8 from a wrecked stock car and scavenged aluminum cylinder heads an intake mounting two Stromberg carburetors allowed Al to stave off the Chevy onslaught on the streets of Philadelphia for a time. The successful installation of an "Isky (Iskendarian) 3/4 race" camshaft, adjustable valve lifters, and 4.11 rear end gears provided Al with the pride of accomplishment and the ability to continue winning street races that fed his competitive spirit.

A common practice among those young men addicted to all things automotive at the time was to obtain car-related publications and digest every word written about performance that could be found therein. Al Joniec was no exception. He quickly decided that the articles written by Roger Huntington, a prolific contributor in many of the era's car magazines, spoke to him in a way that he could understand and learn from.

NATIONWIDE CAR CULTURE

Anytime the history of the car culture, and particularly hot rods are discussed, the state of California takes center stage, even though the major car manufacturers of the mid-20th century were all located near the so-called Motor City, in Detroit. California, particularly the southern part of the state, had several

The acquisition of a modified Flathead V-8 from a stock car provided Al with some top-of-the-line speed parts, such as Offenhauser cylinder heads and an Edelbrock intake manifold mounting twin carburetors. These pieces, coupled with an Isky 3/4 race camshaft completed the package and turned Al Joniec's '40 into his first winning Ford.

advantages. First was the year-round fair weather conditions that encouraged outdoor activities conducive to racing events. Second, a majority of the companies engaged in the manufacture of speed parts, as well as the publications devoted to the coverage of hot rods, were located on the west coast. A third factor would be the dry lakes that dotted the California landscape. Wide open, hard-packed stretches of ancient lake beds were custom-made for the speed and acceleration contests that were a huge part of the hot rod culture.

This is not to say that major metropolitan areas like Philadelphia, which Al Joniec and countless other automobile enthusiasts called home, were mere backwaters when it came to hot rods. By the mid- to late 1950s, one would be hard-pressed to find any location across America, from the biggest cities to the smallest towns, that didn't have a gas station, hamburger stand, diner, or drive-in where teenagers could be found congregating with their cars. Invariably a quiet, straight stretch of road could be found nearby, where two young men could settle any disputes regarding who possessed the quicker car. On any given day, in garages, driveways, and back alleys from coast to coast, young men could be found engaged in the modification of the machines that not only gave them freedom but were also a large part of their individual identities. While California would become best known for the drag race vehicles classified as Dragsters and Gassers, the Northeast would later become the mecca for Super Stocks, the drag cars that most resembled cars that could be purchased off the showroom floor at your local dealer. Al Joniec was destined to become one of America's top Super Stock racers.

In 1954 Al's friend Pudgy bought a brand new 1954 Ford Skyliner (a stylish two door hardtop with a tinted Plexiglas insert in the top over the front seat), powered by the first overhead valve V-8 engine in the Ford line. The two young men washed and waxed the car incessantly, after which they would cruise the streets of Philadelphia looking to meet girls and/or find other cars to race against. They also set about modifying the Ford to make it stand apart from others and enhance its performance. The addition of a set of straight-through mufflers ensured that the duo would be noticed. A device called a mini supercharger installed on the engine did little to add power, but it did create a backfire out the exhaust upon deceleration that the lads found pleasing so it stayed.

THAT ONE GUY

I'm positive that those who grew up in and around the car culture of the 1950s and 60s will recall that one guy. You know the one. He had the coolest car, the fastest car, the nicest looking girlfriend. For those readers not old enough to remember these times, the character known as "The Fonz," on the television series *Happy Days* will assist in forming a fairly accurate visual image. And of course there were the nicknames that invariably accompanied the cool guy.

In and around Philadelphia, and particularly around the street racing spot called the Meadows, there was Joe Fud (Joseph Fudareri was his given name, as Al recalls). Joe Fud was one of the founding members of the Philadelphia Modifiers car club (which remains in existence to this day). He drove a full custom 1950 Ford convertible with a padded Carson top, exhausts that exited through a custom-made rear bumper, and a hand-rubbed black lacquer paint job that was so "deep" it appeared to be liquid. Al remembers being in awe of Joe Fud and his cool Ford.

But it didn't end there. Along with the cool car Joe Fud had a job at THE location for all things relating to hot rods, Frank Bash's Speed Shop in Germantown (a section of the city of Philadelphia). Obviously Joe Fud respected the 16-year-old Joniec enough to get him a job working at Bash's shop doing real high-performance work on customers' cars. Bash was the area's only dealer for the McCulloch supercharger, and it wasn't long before Al was installing them on cars as part of his duties in the shop. An early recollection of Al's was questioning why the McCulloch unit always required the installation of a box over the carburetor to seal it off from the outside. It would be some years later that he learned of the need for the carburetor's fuel bowl to be pressurized in order to compensate for the effect of the supercharger. Al's mechanically inquisitive mind would lead to his own performance enhancing innovations in the future.

While the dream job at Bash's Speed Shop certainly started Al Joniec's climb up the ladder in the high-performance world, his employment there ended on a less-than-positive note. One day while using what was deemed an inappropriate tool to remove a frozen wheel from the front of a customer's '40 Ford, Frank Bash loudly berated his young employee and fired him on the spot. Other employees, who had recommended the tool in question as being the best solution for the job, stood by silently, apparently in fear of losing their own jobs to the sometimes cantankerous Mr. Bash. Al gained a lesson in workplace politics at Bash's, in addition to the mechanical know-how that was one more step in Al's journey to national prominence as a drag racer.

By 1957 around Al Joniec's social circle, anyone who could afford one, owned a '57 Chevy. Stanley "Stosh" Mariannski was one of the lucky ones; his '57 Chevy was light years ahead of the beat-up '40 Ford coupe that Al drove, so young Joniec was quick to accept when Stosh asked him to take the Chevy in for service at the local dealer. After picking the car up from the dealership Al thought a drive out to the Meadows was in order. A guy known as "Mambo" happened to be there with his salmon and white '57 Chevy, looking for a race. In spite of never having driven Stanley's car in a race, Al was quick to accept the challenge. While Al was able to defeat Mambo, it quickly became apparent that the car's anemic three-speed column-shifted manual transmission was not up to the task of competition, as the second gear synchronizer did not survive the encounter. (Chevrolet dealers across the land were inundated with warranty claims on transmissions broken by heavy-footed hot rodders). Thankfully Stanley got over his initial anger over the damaged transmission, figuring the synchronizer was a small price to pay for the victory over Mambo's similar Chevy..

Al's local reputation for tuning and modifying the Chevys grew with every victory in the street races at the Meadows. Cars with the Joniec tune continued their winning ways at Hatfield Speedway, a half-mile oval track outside of Philadelphia, where 1/5th-mile drag races were run on the track's front straight. As Pudgy had the financial wherewithal to underwrite Al's mechanical talents, his Chevy received modifications that others could only read about in car magazines. The 283 V-8 in Pudgy's car was removed, bored and stroked to 327 cubic inches, and topped with a McCulloch supercharger; a Corvette four-speed transmission and 4.56 rear-end gears completed the package. The Chevy was not

In 1958 Ford followed up its successful 1957 sales year with a complete redesign of all their models. Quad head- and tail lamps set the new Fords apart from their predecessors visually. Perhaps the most visually appealing of Ford's 1958 models was the Fairlane Sunliner convertible in Raven Black. It was just such a car that caught the attention of Al Joniec.

Ford introduced the FE (Ford-Edsel) engine series in 332 and 352 cubic inch versions, aimed at replacing the 312 cubic inch Y-Block V-8. While the Y-block would continue to be offered until 1962 in Ford passenger cars, it was limited to 292 cubic inches of displacement after the FE arrived. The FE series didn't exactly get off to a great start performance-wise in its first two years of production thanks to Ford's strict adherence to the AMA ban on factory high-performance vehicles and racing participation. While the Interceptor Special version of the new 352 boasted a factory rating of 300 horsepower, it did not prove to be a startling performer.

built for any specific class at the track but was nonetheless king of the streets around Philadelphia.

The place to go, be seen, and find street races around the area at that time was Bob's Big Boy, a hamburger joint on Roosevelt Boulevard in Northeast Philadelphia. Everyone who was anyone among the local street racers could be found at Bob's. Al would be on hand with the various quick Chevys he had a hand in preparing. When a street race was negotiated, the parking lot would empty out and head to the race location.

Al Joniec's life would take another important turn one day at that Bob's Big Boy. A young man named Louie Tomaselli and some friends cruised in with a brand-new, black 1958 Ford convertible. Louie was just 15, so he had a friend handling the driving chores. Al recalls some of the other regulars at Bob's Big Boy making derogatory comments about the new Ford (the Chevy vs. Ford rivalry goes back a long way and continues to this day), but Al was taken by the new Ford's style and thought it looked great. Louie's car was powered by the newly introduced, 352 cubic inch FE "Police Interceptor Special" engine rated at a very liberal 300 horsepower.

Louie's friends had been trying to help him win his first drag racing trophy. They had had installed an Isky camshaft in Ford's engine to no avail—one of Al's friends even beat the new Ford with his stock 283 Chevy in a street race. After finding out that Louie's dad was offering the princely sum of $100 to anyone who could help his son win that elusive first trophy, Al was quick to offer to give it a try. Now 18 years old, Joniec had ample experience making Chevys run. With his experienced ear he listened to the Ford's exhaust note and determined that the camshaft in Louie's engine had a lobe wiped off, causing it to be way down on power. As it turned out, when the Isky cam was installed the original hydraulic valve lifters had been re-used, a major mistake that caused the failure. Using the most rudimentary of tools (he still didn't own a ratchet set) Joniec dove right into the new Ford's engine. He even used a 2x4 and a hammer to remove the vibration damper—specialized pullers for this type of job were yet another luxury he couldn't afford at the time. With the camshaft removed Louie and his dad were provided with proof of the accuracy of young Joniec's diagnosis. Al got the green light to obtain whatever parts were needed to make the Ford a winner. Al learned that the first 20,000 FE engines produced came equipped with solid valve lifters, so he ordered up the parts required to make the swap from a local Ford dealer. With the installation of a 4.11 rear, the new Ford was ready, willing, and able to take on the local Chevys on the street. By Al's account, Louie never lost another race.

The duo took the Ford out to Hatfield Speedway and in short order returned home with Louie's first trophy. Soon the $100 promised by Louie's dad was in the hands of a young man who would go on to put many more Fords in the winner's circle.

Louie's father was the owner of Jean's Hoagies, at the time the largest establishment of its kind in Philadelphia. Jean's Hoagies was second in popularity only to Bob's Big Boy among young local hot rodders. Being financially well off and respected in the community, Louie's dad had enough influence to have the local police shut off traffic in front of his store. Al, Louie, and their friends could then street race there, using the red light at the corner as a starting point.

Louie's Ford, with Joniec driving, continued its winning ways in organized races as well. They entered the big Winter Race at Hatfield Speedway in 1958, where the winner of Top Stock Eliminator would receive an expenses-paid trip to the Winter Drags in Spruce Creek, Florida. The facility at Hatfield was full of cars and spectators in spite of the freezing weather that day. Louie's dad, coatless and still wearing his white shirt, pants, and apron from work, was in attendance to cheer the boys on. Al expertly guided the '58 Ford through round after round of competition to claim the A/Stock class victory and entered the Top Stock Eliminator runoffs with confidence. As day turned to evening, one competitor after another fell before the Ford until there was but one car left to defeat.

Unbeknownst to Joniec, in the round before the final runoff a car had spilled water on the starting line, which was located at the exit to one of the circle track's banked turns. The water had frozen quickly in the cold air. The Ford was staged in the lane where the ice was forming. When Al released the clutch on the starter's signal the car immediately went sideways. Al recovered and ran down the other car, but the race appeared too close to call. The win was ultimately given to the other competitor. As the crestfallen Joniec drove back toward the starting

This faded photo shows Al's friend Louie posing proudly with his 1958 Ford, the car that would take Al Joniec away from his Philadelphia home for the first time and launch his official drag racing career. The car's Raven Black factory paint job was enhanced with gold scalloping and the name "Hoagie," in deference to the sandwich shop Louie's dad owned.

Al Joinec (left), Louie Tomaselli (right) and a third friend pose with the first trophy claimed by Louie's Ford. (Tomaselli family photo)

line, he was shocked to see Louie's dad engaged in a verbal exchange with the track officials, whom he loudly characterized as, "Goddamn dumb heads," in an attempt to have the race re-run in the spirit of fair play. However, the somewhat dubious winner was a regular at the track, and there was no way the race would be re-run. The official decision stood.

Not willing to accept this injustice, Louie's dad announced that he would sponsor the boys' trip to Florida. This would be a great adventure for a young man who had never ventured very far from the Philadelphia area, but Al was up to the challenge. After swapping out the 4.11 rear end for a more highway-friendly 2.91 ratio, he and Louie set off for their first big-time racing experience. Interstate 95 had not even been thought of yet, so the trip south required traveling on Rt. 301. This road passed through many small towns along the way. More than once they came across someone at a local stoplight who wanted to see just what the out-of-town Ford had. Knowing that the 2.91 rear would be a detriment to the clutch in any races from a standing start, Al and Louie worked hard to convince would-be challengers to race from a rolling start. Al describes the results as being like "taking candy from a baby," as even new Corvettes fell victim to Louie's Ford. Thanks to the tall rear end gearing the Ford was capable of pulling 60 mph out of first and 90 mph out of second gear with Louie yelling "Seeee Yaaa !" out the passenger side window.

The original plan had been to drive straight through to Spruce Creek, but somewhere in Florida in the middle of the night the car ran out of gas. They coasted into a nearby parking lot and decided to sleep in the car until morning. Came the dawn and the young men awoke to the sound of laughter and people staring into the car at them. It turned out they had rolled into the parking lot of Marine World. Soon the boys found themselves attending their first ever seal and dolphin show—quite a unique experience for two kids from Philadelphia area in the late 1950s.

Upon their arrival at Spruce Creek, the boys found that it was nothing more than an old WWII emergency runway stuck way out in the middle of some very swampy terrain. The races were run at night and the only lighting was provided by two giant surplus WWII spotlights that pointed down the track. The enormous

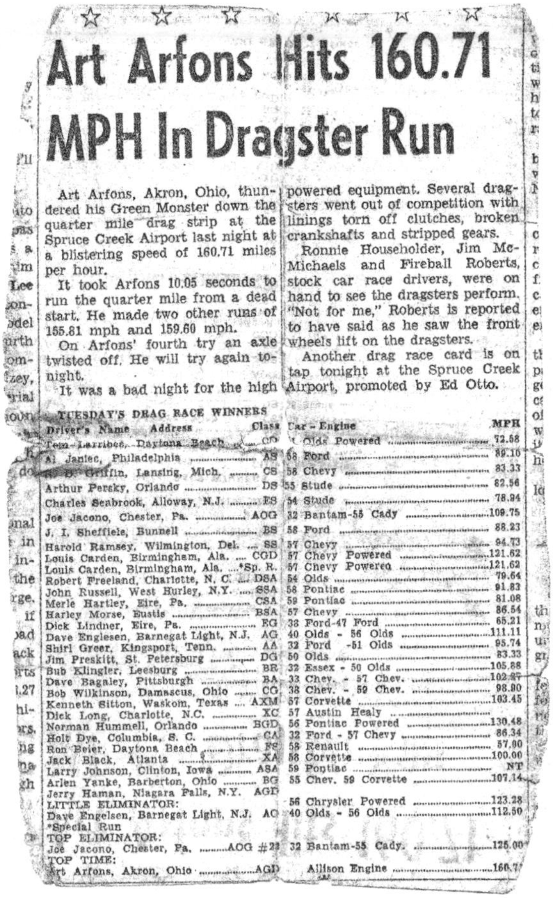

Art Arfons Hits 160.71 MPH In Dragster Run

Art Arfons, Akron, Ohio, thundered his Green Monster down the quarter mile drag strip at the Spruce Creek Airport last night at a blistering speed of 160.71 miles per hour.

It took Arfons 10.05 seconds to run the quarter mile from a dead start. He made two other runs of 155.81 mph and 159.60 mph.

On Arfons' fourth try an axle twisted off. He will try again tonight.

It was a bad night for the high powered equipment. Several dragsters went out of competition with linings torn off clutches, broken crankshafts and stripped gears.

Ronnie Householder, Jim McMichaels and Fireball Roberts, stock car race drivers, were on hand to see the dragsters perform. "Not for me," Roberts is reported to have said as he saw the front wheels lift on the dragsters.

Another drag race card is on tap tonight at the Spruce Creek Airport, promoted by Ed Otto.

TUESDAY'S DRAG RACE WINNERS

Driver's Name Address	Class	Car - Engine	MPH
Tom Larribee, Daytona Beach	CO	[illegible] Olds Powered	72.58
Al Janiec, Philadelphia	AS	58 Ford	89.10
[illegible] Griffin, Lansing, Mich.	CS	58 Chevy	83.33
Arthur Persky, Orlando	DS	55 Stude	82.56
Charles Seabrook, Alloway, N.J.	ES	54 Stude	78.94
Joe Jacono, Chester, Pa.	AOG	32 Bantam-55 Cady	109.75
J. I. Sheffiele, Bunnell	BS	58 Ford	88.23
Harold Ramsey, Wilmington, Del.	SS	57 Chevy	94.73
Louis Carden, Birmingham, Ala.	CGD	57 Chevy Powered	121.62
Louis Carden, Birmingham, Ala.	*Sp. R.	57 Chevy Powered	121.62
Robert Freeland, Charlotte, N. C.	DSA	54 Olds	79.64
John Russell, West Hurley, N.Y.	SSA	58 Pontiac	91.83
Merle Hartley, Eire, Pa.	CSA	59 Pontiac	81.08
Harley Morse, Eustis	BSA	57 Chevy	86.54
Dick Lindner, Eire, Pa.	EG	38 Ford-47 Ford	65.21
Dave Englesen, Barnegat Light, N.J.	AG	40 Olds - 56 Olds	111.11
Shirl Greer, Kingsport, Tenn.	AA	32 Ford -51 Olds	95.74
Jim Preskitt, St. Petersburg	DG	50 Olds	83.33
Bub Klingler, Leesburg	BR	32 Essex - 50 Olds	105.88
Dave Bagaley, Pittsburgh	BA	33 Chev. - 57 Chev.	102.27
Bob Wilkinson, Damascus, Ohio	CG	38 Chev. - 59 Chev.	98.90
Kenneth Sitton, Waskom, Texas	AXM	57 Corvette	103.45
Dick Long, Charlotte, N.C.	XC	57 Austin Healy	
Norman Hummell, Orlando	BGD	56 Pontiac Powered	130.48
Holt Dye, Columbia, S. C.	CA	32 Ford - 57 Chevy	86.34
Ron Beier, Daytona Beach	FS	58 Renault	57.00
Jack Black, Atlanta	XA	58 Corvette	100.00
Larry Johnson, Clinton, Iowa	ASA	59 Pontiac	NT
Arlen Yanke, Barberton, Ohio	BG	55 Chev. 59 Corvette	107.14
Jerry Haman, Niagara Falls, N.Y.	AGD	56 Chrysler Powered	123.28
LITTLE ELIMINATOR:			
Dave Engelsen, Barnegat Light, N.J.	AO	40 Olds - 56 Olds	112.50
*Special Run			
TOP ELIMINATOR:			
Joe Jacono, Chester, Pa.	AOG #23	32 Bantam-55 Cady.	125.00
TOP TIME:			
Art Arfons, Akron, Ohio	AGD	Allison Engine	160.71

This 1959 clipping from a Daytona, Florida newspaper lists the A/Stock class winner at the Winter Drag Racing Championship in Spruce Creek as Al Janiec (spelled incorrectly, as was often the case) driving Louie Tomaselli's 1958 Ford. This was quite possibly the first national drag racing event won by a Ford powered with the new FE engine, but it certainly wouldn't be the last. Al would see to that.

lights blinded drivers on the return road. The lack of creature comforts was offset by the sheer number and quality of the cars entered in the event, including top racers like Art Arfons with his *Green Monster* jet-powered dragster. The *Green Monster* could top 160 mph in the quarter-mile, an incredible performance for the time.

After swapping out the 2.91 rear for the 4.11 and making some timed runs, Al was quick to realize that the amount of dirt and lack of rubber buildup on the starting line was creating traction problems. The Ford was running with street tires (narrow bias ply tires on 14" rims), as required by the class rules. To remedy the situation they re-installed the 2.91 rear.

This proved to be a smart call as Al wheeled the Ford through round after round of competition (he estimates there were 50 cars entered in the class). Al won the A/Stock class with a trap speed of 89.10 mph with the car still in second gear due to the 2.91 rear axle ratio. Al points out that there were no elapsed time clocks at the track and the start of each race was initiated by a flag man.

This was stock-class drag racing in its purest form. The cars were indeed the same vehicles driven daily by the racers. Louie's Ford had the factory column-mounted lever to shift gears in the three-speed manual transmission, (affectionately known as "three on the tree"), did not have tubular steel exhaust headers or even open exhausts, and raced on the same tires it drove in on from Pennsylvania.

Al firmly believes that his class victory at the Winter Drags may well be the first nationally recognized drag racing event win by a Ford powered by the FE series engine, but of course it would not be the last. Al fondly remembers visiting Daytona Beach on that trip to Florida and the thrill of driving Louie's Ford at speed with the top down on the hard-packed sand, where so many historic races and speed trials had been run, as the salt spray splashed their faces. They were just two teenagers without a care in the world, enjoying an adventure in a simpler time.

A COLOR COMBINATION THAT WOULD RETURN

In keeping with the trends of the time, Louie's '58 Ford was blessed with some custom touches and a proper name. The car's black paint job was accented by gold scalloping. The name "Hoagie" was applied in gold with white pinstriping around it at the top of the driver's door—a nod to the Tomaselli family hoagie shop. The black with gold scallop paint job would again pay dividends for Al Joniec in 1966, as the reader will discover in future chapters.

Returning to Philadelphia flush from their victory at the Winter Drags, the team of Joniec and Tomaselli decided to get more serious about drag racing the Ford. The learning curve was about to get much steeper. One of the most popular classes of drag racing competition at the time that seemed to be within reach was C/Gas. Gas classes were governed by a ratio of cubic inches to weight, with modifications to engines and bodies allowable. In hindsight a 1958 Ford convertible is hardly the image that comes to mind when one thinks of a "Gasser," but the boys had already chosen the class and the Ford would be the car with which to make the effort.

Using the boom of a borrowed tow truck, Al removed the engine from the car and undertook his first ever FE Ford engine build-up. In went high-compression Jahns pistons, an Isky roller camshaft and high rev kit, Edelbrock intake manifold mounting six two-barrel carburetors, and a Roto-Faze ignition system that was so complex for the time that Al describes it as "Frankenstein stuff." With a set of tubular steel exhaust headers courtesy of a local welder/fabricator, 5.38 ratio rear end gears, and a set of 12" wide slicks, Louie's Ford was no longer a daily driver.

The complexities of the ignition and the linkage for the six carburetors caused Al to wonder just what he had gotten himself into. But when the task was completed and the engine fired for the first time any trepidation was quickly erased by the roar emitted from the headers. By this point in time Louie was old enough to drive, so he and Al took the car out to the Meadows for its first test. It was a warm summer night, the convertible top was down, and the Ford's engine was bellowing through the headers as Louie slowly cruised up and down the road to break in the engine.

Al finally requested that Louie stab the throttle from a slow first gear roll in order to test the engine's response. When he did the front end came up and the car accelerated with such force that the seat track broke, sending them both into the back seat. Al describes a feeling of awe come over him when he felt the car's newfound power.

In order for the car to fit into the C/Gas class, a considerable amount of weight would have to be added. The solution was simply to pour approximately 200 pounds of concrete into the area between the quarter panels and trunk. This amount of weight over the rear wheels, combined with the slicks, provided a heretofore unheard-of amount of traction off the starting line. An unfortunate side effect was driveline breakage that the car seemed to suffer on a regular basis. Axles were the first to suffer (Ford would not see the need for heavy duty axles with 31 splines until 1963). Ring and pinion gears, yokes, universal joints, clutch hubs, transmission input shafts, and synchronizers all fell victim along the way. When the bellhousing split the team finally admitted defeat.

One final attempt at making the Ford race—and live—was made at the 1959 NHRA Nationals in Detroit. By this time Louie had bought a new Pontiac that Al had tuned for him to race. They used it to tow the Ford out for the big race. Al's strategy was to baby the car off the starting line and lift his foot from the throttle on each shift to prevent breakage of driveline parts. The car was so powerful that this tactic proved effective. Al easily outdistanced his opponent in the first round of competition. In the second round things were going the same way until a clip fell off the throttle linkage. The engine dropped to an idle in third gear, allowing the other car to win by the slimmest of margins. Al and Louie abandoned their efforts with the '58 Ford out of frustration after that race, though Al would later realize the benefits provided by this experience with the FE Ford engine. It could also be counted as Al's first experience with a sponsored racing effort. Even though he was never paid in cash, he had all the free hoagies he could eat courtesy of Louie's dad, who also picked up the cost of any parts required to keep the fledgling race team going. The good news was that Louie did win his class with the Pontiac tow car.

CHAPTER TWO

THE SWENSON FORD CONNECTION

NINETEEN FIFTY-NINE would prove to be a pivotal year in the life of Al Joniec. He formed relationships and took steps forward that would eventually enhance his reputation as a drag racer to the point where Ford would sign him on as a member of its factory drag team. But in the meantime, there was a lot of hard work and racing to be done.

Another of Al's buddies from his Bob's Big Boy days was Jim Rahm, who was George Hurst's original partner in a two-bay, dirt-floor garage in Ambler, Pennsylvania, before it became the highly successful aftermarket speed equipment business that carries the Hurst name to this day. Jim owned a brand-new 1959 Chevy Impala that he offered to loan to Al in order to take his new bride, Barbara, on their honeymoon to Florida. Al and Jim would go on to share many adventures and remain the best of friends some sixty years later, though the marriage to Barbara was short-lived.

Al's next foray into the world of organized drag racing came late in 1959 when Jim Rahm ordered a 1960 Chevy Impala that was equipped with the 350 horsepower, 348 cubic inch high-performance engine, Chevy's top offering at the time. Jim took the car to the Daytona Beach Speed Trials and shocked the Ford and Chrysler factory teams by posting a terminal speed of 132 mph in the flying mile. Jim named the car *Jack Pot* and also lettered it with the results of his Daytona victory. Jim enlisted Al to handle the car's conversion to a drag racer. Al applied all the tuning tricks he had learned while making Louie's Ford a winner, and it wasn't long before Jim's Chevy was making a name for itself on the street and strip. Jim and Al's nemesis was the 1959 NHRA Nationals-winning '57 Chevy of Harold Ramsey, which was a lightweight 210 model sedan powered by a fuel-injected 283 small block. Ramsey's Chevy was the fastest S/S car in the country, but it wasn't beating them by much. After some analysis Al realized that when Ramsey's lighter car got out in front right off the starting line, Jim had a tendency to floor the throttle too soon, causing the slippery street tires to lose traction.

Having recorded quicker elapsed times in the car than Jim had, Al was tapped to drive against Ramsey and he eventually defeated the national champion. Al fondly recalls the moment when Ramsey, whom he looked up to, congratulated him on his win.

Around the same time Al Swenson Ford on Kensington Avenue in Philadelphia was advertising for mechanics to work the night shift in its service department. Al's friend Pudgy hatched a scheme to get himself and Al hired there, and to get around any question of previous dealership experience—normally considered very important in landing a mechanic's job. A nearby Chevrolet dealership had

Al's best friend Jim Rahm with the trophies he won by posting the highest terminal speed for stock production cars at the 1960 NASCAR Daytona Speed weeks. Jim's 1960 Chevy Impala, powered by a 350 horsepower, 348 cubic inch engine, was sponsored by his employer Degnan Chevrolet of Philadelphia and tuned by Al Joniec. (Jim Rahm)

recently gone out of business so the two would simply list the defunct dealer as a previous employer. Just like that, two city kids with limited tools and no dealership experience went to work for Swenson Ford. The night shift proved to be the best possible position to provide on-the-job experience with minimal supervision. Young Al Joniec prospered thanks to his natural ability and willingness to learn. The time at Swenson Ford would prove to be a big step forward and allowed many doors to open.

Flush with local racing success Jim Rahm decided to take his Chevy to the next level in order to compete at the 1960 NHRA National Drags in Detroit. To this end the engine was pulled, its cylinders were bored the allowable .060" oversize, the rotating assembly was balanced, and a blueprinted camshaft was sourced from GM. Because Jim was working at Degnan Chevrolet just a short distance from Swenson Ford, Al was able to spend time preparing Jim's car for the upcoming event. However, they had no time to spare to break in the engine or test the car before they needed to flat-tow it to Michigan. At one point during the journey Al, who was riding back in the race car, decided on a method to break in the engine. He had Jim put the tow vehicle in neutral as he used the race car to push it along at speeds reaching 100 mph. Needless to say this, "scared the hell out of Jim," according to Al.

After a time they pulled to the side of the Pennsylvania Turnpike and unhooked the tow bar to see how the racecar performed on its own. Thundering along the turnpike in the middle of the night, open exhausts blasting, it became quickly apparent to Al that all their hard work had paid off. The Chevy had what it took to be a winner in Detroit.

However, during time trials at the Nationals Al couldn't understand why the car that had felt so strong was now running so poorly. He tried everything to remedy the situation, including rebuilding the carburetors and replacing the ignition system and valve springs. Nothing helped, and he lost in the very first round of eliminations. Dejected, Al and Jim ran

Ford increased the cubic inch displacement of the FE series engines to 390 for 1961 and continued the successes of the previous year's high-performance 352 engine. One version of the new "Thunderbird High Performance" 390. featured single four-barrel induction and was rated at 375 horsepower, while the other mounted three Holley two-barrel carburetors and carried a rating of 401 horsepower. It was Ford's first multi-carburetor offering since 1957. (Zach Straights)

out of gas while driving the car back to their motel. Al coasted into a gas station where he filled up with high test. Upon leaving the station Al jumped on the gas hard and was startled when the car took off like the rocket it had previously been, spinning the rear tires on each shift. It quickly became apparent that something as simple as a bad tank of fuel had cost them a potential national event victory.

AN INTEREST IN TECHNOLOGY

While at Swenson Ford, Al had shown an interest in some Sun engine diagnostic equipment that appeared to be merely collecting dust in the service department. It seemed that most of the older hands treated the new-for-the-time technology as something akin to voodoo. Joniec was more intrigued than jaded about the diagnostic machine, so he signed himself up for the Sun Diagnostic school. After graduation he became

When Swenson hired a new general manager who was a devotee of racing and performance he and Al became fast friends, traveling to NASCAR races together. It wouldn't be long before this relationship paid dividends for Al with the offer of his first dealer-sponsored racecar.

Ford Motor Company stepped up their performance game in 1962 with the introduction of yet another high-performance, FE-based powerplant boasting 406 cubic inches of displacement. Ford supplied owners of cars powered by the new engine with an informative pamphlet listing its unique parts.

SUPER HIGH PERFORMANCE
6V 405 HP

The super high performance Galaxie engine is the top of Ford's engine line. Never before has Ford offered an engine packed with such power and top flight performance.

With the six-barrel carburetion this engine develops a full 405 horsepower and has been track-tested at over 160 miles per hour. This is truly the engine for the prospect who puts high performance above any other feature in his automobile. Ideal for the man who wants

the in-house diagnostic expert at Swenson Ford. His newfound knowledge provided a unique opportunity when the owner of a Ford dump truck offered Al the princely sum of $200 above and beyond the normal service department fee if he could diagnose an engine problem that had eluded other mechanics. It seemed that the truck's engine (a flathead V-8) suffered from exhaust valve guide failures. The owner said even swapping in a good-running engine had done nothing to stop the problem from recurring. Al removed the intake manifold and sure enough, there was a dislodged exhaust valve guide. Instead of tearing into the engine, though, Al deduced that the cause of the failure was exhaust back pressure.

As the young mechanic removed an exhaust manifold from the engine, the truck's owner assured him that the manifold had been thoroughly cleaned during the engine swap. This proved not to be the case at all. The manifold was clogged with material that built up over a period of time in an area between the ports. With the manifold cleaned and valve guide reinstalled, the engine ran as designed and the owner had egg on his face. He told Al that he wanted to test drive the truck prior to paying him the promised $200, but the man never returned.

Al's reputation for automotive diagnostic expertise attracted an offer of an instructor position at the Spring Garden Institute, a school known for vocational training. The young mechanic found teaching very rewarding, as it forced him to learn the various topics he was tasked to teach through research. Al took a hiatus from teaching to concentrate on his drag racing career, but he eventually returned when sponsorship money dried up in the early 1970s.

Ford Motor Company finally shed the yoke of the American manufacturers' ban on factory high-performance cars in 1960; by 1961 the company was fielding some very competitive cars. Al, however, had become more focused on fatherhood and making the living that went along with that responsibility. He did drive a new 401-horsepower 390 Ford that belonged to the parts manager at Swenson Ford in competition occasionally, taking home a couple of class wins at Atco Dragway in the process.

But 1962 would prove to be a completely different story. Swenson hired a new general manager, a younger man who hailed from the South and was an avid racing fan. He and Al became fast friends. When a brand new '62 Ford Galaxie 500 two-door hardtop was unloaded at the dealership—packing an equally new 405 horsepower, 406 cubic inch high-performance engine—the manager asked Al how he felt about racing the car on behalf of Swenson Ford. Al could barely contain his excitement—he quickly began to prepare it to face the latest offerings from GM and Chrysler on the drag strip in the highly competitive Super Stock class.

Al Joniec and Bud Rubino pose proudly with the Swenson Ford 406 Galaxie. With this car Al competed in the top class for factory performance cars at the time, Super/Stock. Al applied many of the tricks he had learned working on the FE engine in Louie Tomaselli's car to the 406 Galaxie to make the car a regional winner in 1962.

The Swenson Ford Super/Stock Galaxie was dubbed *O'Henry* in deference to the founder of Ford Motor Company. A combination of driving and tuning ability was required to get this powerful, 4,000-pound beast down the track on what amounted to street tires.

Al enlisted friend and fellow Swenson employee Bud Rubino to assist with the car; Bud would remain part of Al's team for years to come. Little could be done in the way of modifications aside from cc'ing the cylinder head combustion chambers, giving the car a performance tune, and experimenting with various types of traction bars in an effort to get the 4,000-pound beast off the starting line. But Al applied the experience gained with Louie Tomaselli's '58 Ford to make the new 406 a more competitive package With each change the car responded with quicker elapsed times. The engine's valve lash settings were tightened and camshaft timing was retarded to give the car more mid-range and top-end power—the engine's rated 405 horsepower was of little use off the starting line on the class-allowable rear tires. In the case of early 60's Super Stock cars, which relied mostly on slightly wider, slightly softer-compound rear tires such as Atlas Bucrons, the journey off the starting line and through first gear was a true test of a driver's skill. An elaborate ballet was performed by the foot controlling the gas and the other controlling the clutch. The ability to slip the clutch while "feathering" the gas was the secret. Joniec's driving skill and persistence in finding the right tuning combination (the car was eventually found to run quickest with the camshaft retarded 8 degrees) turned the Swenson Ford Galaxie, dubbed *O'Henry,* into a winner at local tracks, turning elapsed times as quick as 13 seconds flat in the quarter mile.

Al recalls that he did not lose to any other Ford or Chrysler product during the 1962 season. The only car he could not defeat that season was the lighter '62 Chevy 409 of Dave Strickler, which was prepared by Bill Jenkins.

Early in the 1962 racing season future Mopar factory racer Bob Harrop had received a complete 406 Ford engine as an award for his class win at the Winter Drags in Daytona Beach. He parlayed the hot engine into a short-term sponsorship deal with Rice and Holman Ford in New Jersey, which provided him with a 1962 Galaxie to race. After racing the Ford for a short time, however, Harrop received a more lucrative offer from the Chrysler camp, so the car was offered to Swenson Ford. Al knew that the engine Harrop had installed in the car had limited-production, high-performance exhaust manifolds with longer runners (the type that would become standard equipment on the 427 engines for 1963). The moment that Swenson cut a deal to buy the car, Al set about swapping the exhaust manifolds onto the Swenson Ford Galaxie. In retrospect Al noted that while the long manifolds increased the engine's horsepower as reflected in the terminal speeds recorded at the end of the quarter mile, the weight they added to the front end of the car did nothing to lower elapsed times. When Al learned that a young man from South Bend, Indiana, named Dick Brannan (soon to be named drag racing coordinator at Ford) was stopping the clocks in the high 12-second range with his 406 Ford—a lighter two door sedan version—he wished that he had opted to wait for the lighter model, as 100 fewer pounds equated to approximately one tenth of a second improvement in elapsed time.

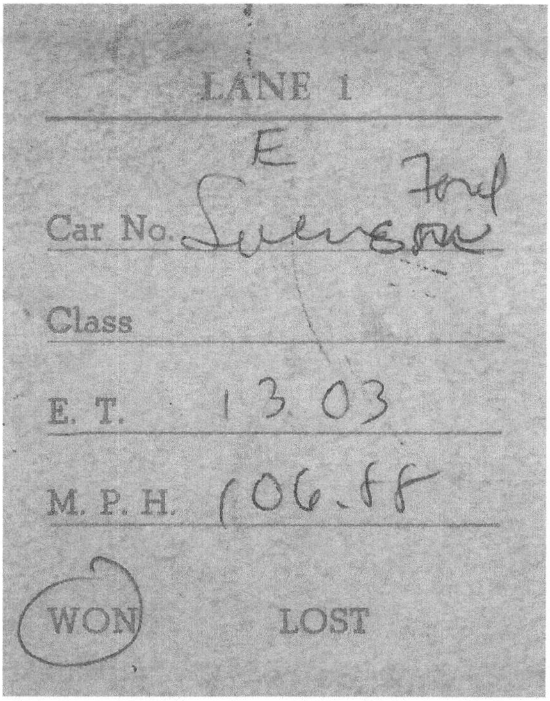
LANE 1

E

Car No. Swenson Ford

Class

E. T. 13.03

M. P. H. 106.88

WON LOST

Al's efforts with the '62 Galaxie paid dividends in the form of very low 13-second quarter-mile elapsed times as can be seen on this time slip. This was right in line with the quickest S/S Fords in the nation at the time, with only Dick Brannan running quicker.

CHAPTER THREE

ATTENTION FROM FORD

THE SPRING OF 1963 brought the highly anticipated arrival of one of the new lightweight Galaxies—offered to dealers for drag racing competition that year—to Swenson Ford. While he looked forward to getting his hands on the car, with its 427 cubic inch engine, improved camshaft, and two four-barrel carburetors, Al had two concerns. First, exactly how much would the new car weigh? It was obviously bigger than the previous year's model. Secondly these special cars were being made available to all Ford dealers, so it was not a guaranteed winner. Or, as Al put it, "You had to work your ass off to make it a winner."

As it turned out the car was several hundred pounds lighter than the 1962 Galaxie, thanks to the prominent use of fiberglass and aluminum components and the deletion of any weight-adding creature comforts found in a normal production-line car. After inspecting the car's every nook and cranny, Al secured a dealer license plate from Swenson, rationalizing that he needed to drive the car in order to break in the engine. While some hours of highway operation would serve to loosen up the internal components in the 427 engine, the Galaxie's huge trunk easily accommodated fishing poles and other equipment for a road trip to the mountains.

Upon returning from his fishing trip Al figured, as long as he had the car out, he might as well cruise down to his old street-racing haunt, the Meadows. While a stepped-up police presence had curtailed much of the street-race action at the Meadows, a solo pass proved the car to have great potential. As Al reported, "I did manage to find one poor bastard in some kind of a Chevy that didn't know what I had and I just ran away from him."

Authors note: When it comes to the race preparation of the Swenson Ford lightweight Galaxie, I'll utilize Al's recollections verbatim in the hopes that the reader will get a true feel for the manner in which Super Stock class drag cars were set up for competition in the early 1960s. This is how a racer who didn't have the benefit of factory engineers or computers went about developing a winning car. This, boys and girls, is drag racing the way it was done before money could buy whatever it took to be a winner, including hiring a paid driver.

Satisfied the car could be a big winner in 1963, I made a mental commitment to make a BALLS OUT, BELLY ON THE GROUND attempt to do whatever it took to make it happen. I took the heads off to cc' them and did a classic, thin-seat valve job like I had done in '62. To my surprise the intake seat had a large, flowing taper to the seat radius (what I now call a flow bump). It couldn't be thinned down even if I wanted to without

It took very little time for Al to turn the Swenson Ford lightweight Galaxie into a consistent winner. The coveted title of *Eastern Drag News* Top Dog was bestowed upon Al's new mount.

cutting it out completely. I did thin the valve but not the seat, as I suspected the Ford engineers had put them in for a reason. So I left them alone, but I did do the exhaust seats. Later on, I can't tell you how many guys asked me what I did with the valve seats when they saw that I was really running fast. Literally all of the slower cars cut them out. I believe the unique intake seat radius tricked the air into curling around the seat and valve and aided in low-lift flow. The cams were only around .550" total lift back then. The average valve lift would be low by today's standards, so low-lift flow was very important.

Author's Note: More than five decades later, during the course of a Powerpoint presentation on the history, theory, operation, and performance of automotive engines, Al would explain his previously described deduction and how it ultimately benefitted his performance, even though he didn't understand the physics at the time.

Again, I went through several traction bar configurations. Getting traction with treaded, street-legal Atlas Bucron recapped tires was still a major problem.

For 1963 I was going to get tube headers no matter what, as I had learned my lesson the previous year. I decided to build them myself as I still couldn't buy them. I learned how to gas-weld thin tubing, hunted down U-bends and straight tubing, somehow found flanges, and I began work on my first set of headers. It was a 24-7 effort as the racing weekends didn't wait.

I didn't know how long to make the individual header tubes for proper tuning, so I just ran individual tubes all the way back to the rear axle. This was easy once the nightmare of making the tight bends as equal-length as possible was completed. I used a string with a small nut tied on one end and simply held the header and dropped the nut into the pipe. I used cheap exhaust tubing in 2-foot lengths with slip ends from Pep Boys.

I had pipes up the ass to make it all work that far to the rear axle.

I then went to the Vineland, New Jersey track, which had a wall next to the starting line as it was an oval track and they extended the straightaway (complete with bump) for a full quarter-mile drag race. The wall reflected the exhaust noise back to the car and it sounded wild from the pipes being so long. But I wasn't listening to that, I was listening to my time card. I began to make time trial runs, taking two-foot sections off after each run as the car kept going fastest and faster. I'm still amazed that I wound up with around 38- to 40-inch length primaries—just like today's headers—from the school of hard knocks. The headers worked out for me and really helped the performance of the car. I started kicking ass and taking names that year.

I continued to work on other areas and performed my old trick of closing down the valve lash to as close to zero as I dared. I would put a dial indicator on the rocker and moved the rocker up and down to actually measure the real clearance. I didn't always trust a thin (.003") feeler gauge because there was usually a wear area on the rocker arm tip were the valve stem hit. I always felt that the purpose of valve lash was for heat expansion and machining tolerance when the engine was hot, so why have any extra? Eliminating the valve lash could increase the duration and lift (.015" less clearance was equal to .026" lift).

While doing that one day I put the dial indicator on the rocker shaft at the very ends (all the end rocker arm shafts had no support) and used the starter to turn the engine over. I almost shit my paints as the rocker shafts were moving and/or wiggling around .015" to .025" just on starting motor rpm. I wondered, "what the hell was happening at max rpm, or when a bad resonant frequency started?".

I immediately started looking for solutions to this problem. Realizing that there was a top head bolt on each corner of the head, near the ends of both rocker shafts, I measured the height, both in the center of the shaft and offset of the shafts to the bolts, and figured that a 2- to 3-inch piece of angle iron would take care of business as a support brace. I had the four head bolt tops machined flat and drilled and tapped for a 10/32" bolt. I then took a head to a machinist to have him precision-bore the angle iron for the shaft size hole at the correct location as well as the 10/32" bolt hole. I just had to slip the machined angle iron on the rocker shaft and then got precision thin washers to shim for the 10/32" hole between the angle iron and head bolt. This gave, as near as possible, zero crush or very slight preload on the rocker shaft when tightening them down.

It worked; I now felt a major kick in the ass at the track. Years later I tested it on a dyno and saw a reliable 15- to 25-horsepower gain depending on the cam and valve spring tension. I then went back to retarding the cam again. I had achieved good results with the '62 car when I went as far as 6 degrees retard. This time I wound up with a full 11-degree retard and would have most likely continued further, but the eccentric dowel for the camshaft gear that I was custom-making couldn't take any more offset. I find it interesting that many of today's car engines have the ability to advance and retard their cams a total of 11 degrees each, for a total of 22 degrees. Retarding the cam really worked for me because the stock-type tires we were using meant we couldn't utilize any additional low-end torque. As a matter of fact, I used a close-ratio road racing T-10 transmission with a very high 2.19 ratio first gear as well. This was something I learned from the days with Louie's car. With the close-ratio transmission the engine never dropped more than 2,500 rpm or so between gears. I always revved the engine until it sweated oil and/or just before it floated the valves. You have to be mean to your machine to win races.

After applying these tricks I was going as fast as anybody in the country who was NHRA-legal, according to the drag newspapers. There were guys who match-raced with slicks who went faster, but were not NHRA-legal; that is, except

The first order of business for the Swenson Ford team upon arrival at the 1963 NHRA Nationals was to replace the four-speed transmission in the race car, which was broken due to a mishap while making the long tow from Pennsylvania.

for Dick Brannan, as he was the guy to beat and the best Ford team car. I finally got to race him at the NHRA regional race in York, Pennsylvania, where the Ford team cars of Tasca and Brannan were the cars to beat. Brannan wound up setting the NHRA Super Stock class record and, to my dismay, was in fact a little faster than me that day. After winning some rounds, which included beating Tasca Ford outright, it was now me and Brannan for the trophy. He either underestimated me and/or fell asleep as I put a hole shot on his ass and held him off for the win.

I was happy and relieved. I had not lost to any Ford so far, and now had beaten the official record-holding Ford as well. I knew now that I was more than competitive, so I planned on going to the NHRA Nationals. But I knew that I had to go even faster than before.

Another result of the York win was that I was finally recognized by Ford. I remember I was so nervous talking to Jacque Passino from Ford that my voice trembled. I wanted so badly to be a part of the Ford team. They did send me fiberglass doors, only because I knew that Brannan had them and I didn't (I always wondered what else I didn't have). The doors were lighter and helped a little bit, but then racing was all about a lot of little things all adding up.

I kept running at the local tracks and kept winning. I was now the East Coast S/S "Top Dog" (A highly coveted title bestowed to top performers and record setters by *Eastern Drag News*). The car's name was *Peanuts*, and it only seemed appropriate to have a depiction of Snoopy painted on the trunk lid once top dog status had been achieved."

Al also found his share of match races with the big Ford in 1963. This form of racing, which was steadily gaining popularity among promoter,s pitted two cars, usually of competing brands, against one another in either a best two-of-three or three-of-five runoff. *Mid-Eastern Drag News* reported, "Swenson Ford tops Newcomer Chevy at Vargo Dragway," on August 11, 1963. This best three-of-five match race was scheduled in order that the Chevy might avenge a previous loss to Joniec's Ford that had occurred on July 21.

The Swenson Ford lightweight Galaxie sits in the staging lanes at some long-forgotten East Coast drag strip waiting for the next round of competition late in the 1963 season. After being sold to Lon Razer, the car would continue its winning ways for several more years.

With the match at two rounds to one in favor of the Ford, the Chevy driver jumped the flag twice in an effort to gain an edge off the starting line only to be called back by the starter. When a green flag start was achieved, "Swenson Ford pulled the Chevy out of the chute and the Chevy just couldn't catch the fine-running Ford of Al Joniec as the Ford won the race in three out of four, showing once again that Fords do run." *Eastern Drag News* ran a headline on August 18, 1963, titled "Al Swenson Cleans House at Cecil County." Writer Tommy Lynn reported the action:

> It looked as if Al Swenson Ford and Hall Ford scared off all the competition in Super Stock here at Cecil County today. These two fiberglass Fords battled it out for S/S class, the Top Dog title (and the $50.00 bond that went with it) plus Top Stock. Al Joniec stroked the Swenson Ford to all of the above wins and in winning Top Stock he put down the likes of Harold Ramsey, *Bounty Hunter* and finally Bob Boyd in the Hall Ford. The big white Ford was turning 12.30 and 116 all day.

SO AL, WHAT DO YOU THINK OF THE NEW COBRA?

Thanks to its status as a high-performance dealer, Swenson Ford became eligible to sell the new, very limited production Shelby Cobra, a hybrid British sports car powered by a small-block Ford V-8 that was the brainchild of former racer turned manufacturer Carroll Shelby.

One day Al Swenson told me that they were flying in a brand new Cobra to the LaGuardia (NY) airport. He asked me if I wanted to go pick it up and drive it back. I was taken aback by this, as I only heard about the cars and never saw one, so I didn't know what to say. Here I am racing a full-body 427 Galaxie. Along comes this English shitbox car with a little 289 engine in it that I was supposed to get excited about? I could see his excitement and I didn't want to pee on his parade, so I told him I'd think about it.

New York is 100-plus miles away from Philadelphia and we needed to have two guys to drive to get it, one to drive it back. The logistics were not good. I was racing my car and was very busy, plus I didn't think that much about the car. However, I told Jim Rahm about it and he said, "No problem, I'll fly you there with my brand-new V-tail Beechcraft," as he wanted to put hours on the plane and needed a place to go.

I wanted to experience flying, and it cut out half the negative problems of the trip as well. Al Swenson gave me the dates and paperwork to go pick up the car. Bud Rubino also wanted to get in on it, so he signed up to go as well—there went the efficient use of employees! But we all piled excitedly into the plane.

The estimated flight time was one hour—it sounded like a piece of cake. Most of the flight was smooth and peaceful just like we anticipated it would be, and we were all thrilled with the adventure so far. But as we approached LaGuardia the SHIT HIT THE FAN. The landing approach path was crowded with about six big Boeing 707 jets, at least 2 ahead of us and the rest behind us. The turbulence kicked up by the big jets in front of us was simply unreal. I had one of my hands on the roof and the other on the armrest handle and was screaming at the top of my lungs get us the hell out of there. Jim turned around and said "Look, they are right behind us too."

We were bouncing around like a cork in a violent sea, and I get seasick really easily. I was terrified and so was Bud. I thought the airframe couldn't possibly take it and would break apart. Jim had both hands on the steering yoke with a death grip, trying to keep us in the glide slope. He also had to firewall the throttle to keep up the speed of the big jets, while talking to the control tower. He had his hands full, so I shut up. After a painfully long (but most likely short) time bouncing around we finally got

MID-EASTERN DRAG NEWS 9

WATCH FOR YOUR FAVORITES AT—

Cecil County Drag-O-Way

"The East's Most Ultra Modern Plant"

SUPER TRACTION FOR SUPER ACTION

LOCATION: Just over the Maryland line on route 272, three and one half miles off Route 40 between Baltimore and Wilmington.

PITS OPEN 9 A.M. — TIME TRIALS 12 NOON – ELIMINATIONS 2 P.M.

OCTOBER 13

OLD RELIABLE VS ARLEN VANKE

1963 National Champ in A/FX Old Reliable takes on the present A/FX National Record Holder Arlen Vanke, in his 63 Tempest record holder at 11.89, 123.28 mph. A best 3 of 5 between the Worlds Fastest A/FXers.

PLUS-MEDN "TOPDOG" SUPER STOCK RUN-OFF. AL SWENSON, No. 1 Super Stock in the East VS the Challenger NORRISTOWN FORD.

THE NATION'S FASTEST B-GASSER RETURNS BY POPULAR DEMAND

OCT 20 HASSEL — OCT 20 VOGELSONG

M.E.D.N. B/GAS TOP DOG RUN OFF

EAST COAST QUICKEST B-GASSERS WILL BATTLE IT OUT FOR THE B-GAS SPOT IN MID-EASTERN DRAG NEWS (THE WINNER) RUNS B-GAS NATIONAL CHAMPION, RON HASSEL (BEST 3 OF 5)

WEEKLY ELIMINATORS

TOP ELIMINATOR	— Less then 9 sec. to 9.50 sec.	$50.00 Bond
MIDDLE ELIMINATOR	— 9.50 sec. to 11.00 sec.	$25.00 Bond
JUNIOR ELIMINATOR	— 11.01 sec. to 12.00 sec.	$25.00 Bond
LITTLE ELIMINATOR	— 12.00 sec. or more	$25.00 Bond
TOP STOCK	— All stocks & F/X cars running street tires turning 13.50 sec. and under (you need not win your class)	$50.00 Bond
HANDICAP STOCK	— (No super stocks, F/X or cars running for Top Stock permitted) class winners only.	$25.00 Bond

NOTICE: CECIL COUNTY WILL REMAIN OPEN TO BRING YOU THE FINEST IN DRAG RACING AS LONG AS WEATHER PERMITS.

This *Mid-Eastern Drag News* ad for Cecil County Dragway announces an upcoming "Top Dog" Super/Stock runoff between Al Joniec driving the "No. 1 Super Stock in the East" versus the Norristown Ford lightweight Galaxie near the end of the 1963 racing season.

permission to land directly behind the jet in front of us. I thought for sure the turbulence would stop when we reached the ground, but NO! We continued to bounce around on the ground as well. Only when we taxied away from the big jets did things settle down. Jim did a masterful job of landing the plane in such a stressful situation and became a SKY GOD to me ever since. You have to remember this was in 1963; I'm sure you cannot take a small, private plane and land in the same flight path with a commercial jet today.

Jim took off to go home and so we found ourselves at this huge airport with no idea where to go find the car. It was about 2 a.m. and we were both tired, but we finally found

In this photo taken at the 1963 NHRA Nationals, Al Joniec in the Swenson Ford lightweight Galaxie faces off against the A/Stock Ford of Welsh Motors in a time run. (Snyder Specialty Cars)

the place. Once we got to the right department the rest was easy. The car, a beautiful dark green convertible, was sitting there waiting for me to pick it up. I was impressed. I got in it and started it up and it sounded great. I said, "Hell yeah!" and off we went.

It took us a while to get back to the New Jersey turnpike and it was now going on 3 a.m. Nobody else was on this road, made up of four separate 3-lane sections, all separated by guard rails. Bud went to sleep; the loud exhaust and the wind noise from all the window and door leaks were all you could hear, so even if he was awake we couldn't have talked. I too was tired and wanted to get home and go to sleep as soon as possible. I started to kick up the speed. Around 100 mph the car actually started to get quieter, as the air pressure started to build up in front of the car, forcing air away from the body. I edged up to around 110 but Bud didn't wake up, so I went faster. I was soon passing 120 mph. It was weird as the car seemed to be getting quieter. The exhaust was the only thing that got a little louder at 125.

The turnpike was huge. It was relatively new and smooth, and I could see at least a mile ahead. It was clear so it desensitized me to the speed. I was starting to change my mind about the car, as it was performing beautifully and it was rock steady at this speed. Life was so good, I started to wonder how fast it could really go. As the speed started to climb over 130 mph, all of a sudden BOOM! It sounded like a bomb went off.

I could hear Bud's screaming (he must have dreamt we were going off a cliff or something) above the wind noise, (which he denied later on). I wondered, what the hell just happened? I was as shocked by all the noise as Bud and liked to shit my pants. After the huge BOOM, there was also a big temperature change that slapped us in the face. It was freezing out and we had been comfortably warm and happy inside. Talk about a wakeup call—I was no longer tired!

Turned out the high speed built up enough negative pressure above the convertible top that the rivets that secured it to the soft aluminum body simply pulled out The top was just hanging on by the windshield clamps. This caused an over-pressure inside the car, so when it popped the pressure differential caused the big noise, and we were in the center of it. I pulled off to the side of the road and pushed the top back into place the best I could. I then drove real slow the rest of the way home. The next day I gently tapped down the little volcanos in the aluminum where the rivets had pulled out. I then took each individual rivet and tapped them down into the original holes as best I could (they were still loose) and put the top back down over them. I drove to Swenson Ford and said how much I liked the car (you should never judge before you try). Everyone, including me, thought it looked great. My only complaint to them was a quality control issue, as I noticed that the rivets that hold the top on appeared to be loose."

I planned to go to the Indy race, but up until that time I still hadn't taken the engine out, though I had the heads and front cover off for cam timing changes and so on. In preparation for the NHRA Nationals I took the leap and did a full blueprint of the engine, complete with a .030" overbore. I wanted those extra 3 cubic inches, as everything helps at that level of play.

The one thing I did wrong was not getting a blueprinted cam (you could only use a stock cam). The NHRA officials checked the specs at open and closing degrees and max lift,. Many years later Dick Brannan told me he was using a so-called blueprinted cam with the correct open and closing degrees and max lift, but with much bigger lobe ramps (think of it as more of square lobe instead of a pointed one). He said it was worth at least 25 horsepower.

Aside from the Winternationals held in Pomona, California, the NHRA Nationals in Indianapolis was then the only other really big race of the year. Indy was hundreds of miles away and was the farthest place we ever went flat-towing the car.

Because this was such a long ride and I was always tired from working on the car all the time, someone else usually drove and I slept. This time the tow car was so full there was no room to sleep, so I slept in the race car, which had all kinds of room. Somewhere in Ohio I woke up to the sound of the engine turning over. Confused, I tried putting the clutch in and the engine stopped turning over, so I knew it was a problem in the transmission. We towed it with the drive shaft in. Because we had never towed the car hundreds of miles before, I never realized the output shaft was spinning inside high gear. It wasn't getting any oil due because the cluster gear wasn't turning in the oil, so after hours of spinning without lubrication it froze the high gear to the output shaft. This little surprise had to be fixed before we could even go through tech inspection.

It appeared that every lightweight Ford in the country there, with a very few Mopars and no Chevys that I can remember. Chevy did have a lightweight car (with an aluminum front end), but they didn't make enough of them to be eligible to run in S/S. They had to run M/P (modified production) and/or factory experimental. It was easily the hottest race I had ever been to, with 90-degree temperatures and very high humidity. It was miserable. Most of the teams where not working on their cars like they normally do, but instead just trying to stay cool. I had learned about leaning out carburetor jets to adjust for hot weather by reading a Roger Huntington article, so I started to do that. I watched my time card as I kept putting leaner jets in after every time trial until I was a full 10 steps leaner (though I was most likely to rich to start with). As a result, I was now the fastest car in S/S at the Nationals. Back then there was no computer

A look down at the staging lanes during Super/Stock class eliminations at the 1963 NHRA Nationals shows the Swenson Ford lightweight Galaxie to the left, waiting to be called to the starting line. (Snyder Specialty cars)

posting of any kind, so I gave one of the guys a pen and paper and sat him near a loudspeaker to record the S/S times during time trials—that was the only way I knew I was the fastest (at least in time trials).

AL MEETS ROGER HUNTINGTON

While at the Nationals Al was approached by a young man who told him that Roger Huntington wished to meet him. "That almost knocked my socks off," Al said. "I knew that Roger kept on top of current trends and the Super Stock class was one that piqued his interest. Perhaps he wanted to meet the fastest."

Shortly thereafter the young man returned, pushing a man in a wheelchair, who was introduced to Al as Roger Huntington. Al describes being taken aback by the fact that one of his heroes, a man who he envisioned as a master mechanic/engineer and hot rodder from the stories he wrote, was not even capable of driving an automobile. Al was stunned and thrilled, though, particularly by the fact that the man whose technical articles he had been reading since 1955 had wanted to meet him.

Did I tell you it was HOT? Some teams started using ice to cool both themselves and their engines. I thought that was a great idea—hell, I never saw an engine being iced before. I told the guys to get tons of ice and start putting it on the intake manifold after each run. In eliminations I made it through a few rounds before I faced a Dodge, of all things. Back in those days we had to run street-legal treaded tires, so I fast idled off the line and throttled the engine though first gear. But when I finally floored it, the engine stalled from being too cold. I can still remember I yelled as loud as I could, "NOOOooooo !" I put the clutch in and revved the engine and had caught and passed him by 3rd gear. However, by then, the huge amount of ice that was on my intake started to melt. Water started to fall off the engine and got under my rear wheels. As I started spinning and going sideways in 4th gear, causing me to lift off the throttle, he passed me back. I straightened it out and caught him yet again at the lights. The race was so close I didn't know who won until I got my time slip. Needless to say, my guys had gone overboard and totally packed the intake with ice, which would have been OK if a dragster hadn't blown up just before we went out. During the 20 or so minutes it took to clean it up the track, my intake/engine got totally cold and we were the very next cars to go out. Remember also that I had been leaning out the carb jet, which compounded the problem caused by the over-cooled intake.

I was stunned that this Dodge not only beat me, but beat everyone else and won the whole race. I just couldn't believe it—I know I could have easily beat him if I didn't use ice, and I still relive it, even today. In the end the Dodge was disqualified during the required teardown for having an illegal camshaft and nobody was credited with the S/S class win that year.

After the trauma of losing Indy for the third time set in, I at least had the self-satisfaction that I had the fastest Ford at a race where everyone of importance had been, so now Ford knew I was a player. Shortly after going home Dick Brannan (the Ford Drag Team leader) sent me a fresh-air intake package for my car. WOW—I couldn't believe what a difference it made! It was good for at least 2 to 3 tenths and 3 to 4 mph. It consisted of air cleaner top and bottom, but instead of an air cleaner element, it had a solid metal oval insert with a side connector to clamp on a 6-inch hose that sealed and ducted outside air into the air cleaner. It also had a large, tapered, fiberglass-molded duct that was shaped like a tapered horn and bolted to the radiator support. This required a large cut-out area for letting in the cooler, high-pressure air right from the grille directly to the engine, which in my opinion was as good as, if not better than, a hood scoop. Every 10 degree drop in air temp equaled a one percent power increase. At speeds over 60 mph the new intake created a ram-air effect, which increased as you went faster and acted like a supercharger with at least 1 pound of boost.

Near the end of the season I had to sell the '63 car if I wanted to get a '64. I had a buyer lined up and I arranged to give him the car at Cecil County Dragway in Maryland after the big end-of-the-year Super Stock race. There must have been 50-plus cars there that day, including everyone important from the East Coast. The weather conditions were cool and perfect to go fast. Bill Jenkins was now working on and driving one of the two Bud Faubel 426 Max Wedge Dodges and they proved to be the cars to beat. The old girl was flying that day (12.10s at 120-plus mph) with the new fresh air package, but still S/S-legal.

I was having a fun day, winning round after round, and still remember racing Bill Jenkins in one of the Faubel Honker Dodges. I respected Bill as a tuner and saw that he had both of his cars flying. It gave me great satisfaction to beat him, as when I put it in high gear and turned around to look at him I had pulled away by around two bus lengths (only horsepower does that).

It was now just Bud Faubel and me for the final race. Bud had the faster car of the two Dodges. I had him off the line and was leading him when I put it in 2nd gear and broke the cluster gear in the transmission and lost the race. Though I was disappointed, I had the satisfaction that I could'a/should'a won. The new owner of the car was thrilled that he just bought the fastest legal S/S Ford alive. So I called it a good day and kissed the car goodbye. (Author's note: Lon Raser, the man who bought Al's lightweight Galaxie, also went on to claim the title of *Eastern Drag News* Top Dog in the AA/Stock class for 1964 while winning numerous class victories, eliminator titles, and a number of regional meets in the ensuing years).

Looking back on it now, 1963 was not only a good year for me but a great learning curve as I went from just randomly changing/tinkering with hit-and-miss-type things to strategically tailoring the engine powerband to optimize the available traction and/or gear ratio. As a result I went the entire season without losing to another Ford.

GREAT EXPECTATIONS AND SOME DISAPPOINTMENTS IN 1964

Al and many other Ford racers eagerly anticipated the arrival of Ford's next and most serious attempt to dominate Super Stock class racing for 1964, the

Thunderbolt. It was based on the mid-size Fairlane body style and powered by a version of the 427 called the High Riser, which featured an improved intake manifold and cylinder heads. The new drag race-only vehicles were hand-built by Ford performance subcontractor Dearborn Steel Tubing Company. Taking full advantage of the NHRA's cubic inches to vehicle weight rule in the Super Stock class, Ford would build the required 100 of these special Fairlanes and make them available through the dealer network to serious racers.

The agreement between Ford Motor Company and dealers called for the cars to be picked up directly from Dearborn Steel Tubing's facilities, rather than being delivered via normal new-car carrier. Al Joniec's new mount was delivered to Swenson Ford enclosed in the back of a fruit truck. Unfortunately, the joy of seeing the new car was quickly dashed when Al realized that, in spite of being ordered with a four-speed transmission, his new mount was equipped with an automatic! Al describes his initial reaction as, "Only girls drive automatics." Ford had Dearborn Steel Tubing construct 59 of the 100 Thunderbolts with a modified Lincoln automatic transmission. This was an obvious attempt to counter the dominance of Chrysler products equipped with the superior Torqueflite automatic transmission in the Super Stock Automatic class. Ford pressured dealers to accept the automatic-equipped Thunderbolts, but a great percentage of them would be converted to four speeds as quickly as possible to make the cars more competitive.

Al Joniec is seated behind the wheel of a brand- new, 1964 427 Fairlane Thunderbolt factory Super/Stock race car that has just been delivered to Swenson Ford.

After realizing that exchanging the car for a four-speed version, or obtaining the required parts to convert the car, was not going to happen, Al decided to play the cards he had been dealt. Preparation began with a trip out to The Meadows to see how the car felt. After dispatching several unsuspecting victims in street races, Al began to think that perhaps the automatic wasn't so bad after all.

However, early outings on the drag strip proved that the "race-ready" Thunderbolt was no match for the Chrysler products. Al became depressed when continual efforts to make the car competitive proved fruitless. After explaining his dilemma to dealership owner Alvin Swenson, a telephone call was made to

EASTERN DRAG NEWS 7

FORD LEADS THE PARADE AT ATCO'S $1000 S/S MEET

by Mother Les

ATCO, N.J., June 21—With $1000 (ALL CASH) mind you up for grabs, the super stocks poured into Atco Dragway like it was the closing meet of the season. Of course the $1000 had a little bit to do with it.

Favorite on the betting boards was home town hero, Bob Harrop and his Ram Charger Dodge. Not to be counted out of the picture was Al Swenson's Ford entry driven by Al Joniec. During the first round of eliminations it was these same two cars that were doing the eliminating. Out of the many Ford and Dodge entries Joniec and Harrop were the last to hit the home stretch. With $400 riding on the nose of the winner, both drivers were hot out of the hole heading for pay dirt. It was strickly drivers skill that won this race as Joniec went through the gears a little faster and pushing the Ford Thunderbolt out in front for the lead and the old green stuff. The Ford fans went wild as Joniec hit the Chrondeks at 11.73 and 119.04 mph. Harrop followed it up with an 11.85 and 117.60.

Round No. 2 brought back all the cars again except Al Joniec's Swenson Ford. Putting the axe down on the second round was Harrop again with the help of another Ford entry, that of Bernie Golack of Dobbs Ferry Ford in New York. Second run for the big money with $250.00 up this round, Harrop proved to have learned his lesson putting the Ford Thunderbolt away with a quicker e.t. than that which had beaten him the first round. Compared to the 11.73 of Joniec in the first round, Harrop took the second place money with an 11.68 at 121.13 mph. with Bernie Golack's Ford trailing behind at 11.84 and 117.49.

Final round for the $150 in the third go pitted another Ford, that of Freddie Henderson with his "Chazam" against a Plymouth this time, the '64 automatic of the "Jersey Duke #2" with Jack Worst at the helm. At the GO-GO-GO-GO for the finish line and $150 cash green backs, it was Henderson coming out strong and pulling to a small lead but just enough to put him through the Chrondek ahead of the MO-Par at 11.95 and 118.42.

Joe Gerdelman's Bounty Hunter #1 with Gene Bowman doing the driving chores was top man come Top Eliminator Time, turning the BIG HOG up into the 170's at 8.53.

With no cars eligible for the #3 eliminator spot, #4 consisted of Jack Gormley's Ford powered C/Digger running against the D/Dragster of Charlie Seabrook. The J & J Texaco Station entry proved a little too much for Seabrook's C/Digger by putting the timers on 11.09 at 120.16 mph.

Number 4 eliminator was taken back to New York by Carfi & Gendron in their 6 cylinder C/Gasser. Powered by Chevy and covered by a Henry J body the little C/Gasser did the trick in 12.98 secs. at 108.34.

Top Stock found the age old battle on the line with Arthur Lipkowitz's Chevy vying for honors against Jim Horton's Ford. Chevy must be on their way back because Arthur took the win over the Ford at 13.25 and 109.35 mph.

Next week Atco will sponsor its Annual Powder Puff Derby with the whole spot light going to the ladies. All entries can be made beneath the timing stand, this in itself should be quite a show. Also on this day Atco will be holding a MISS ATCO DRAGWAY BEAUTY CONTEST. So fellows bring your favorite girl or someone else's we don't care. Gates will open at 3:00 pm with Eliminations getting underway at 8:00 pm.

By late spring 1964, Al had turned the Swenson Ford Thunderbolt, now converted to a four-speed transmission, into a winner. This issue of *Eastern Drag News* describes how Al took home a $400 purse by winning a Super/Stock meet at Atco Dragway. Particularly sweet was his victory over archrival Bob Harrop's Dodge in the final.

one of Mr. Swenson's former college friends, who just so happened to be a Vice President at Ford. In very short order all the parts required to convert the car to the desired four-speed configuration arrived at the dealership.

After losing nearly two months of precious development time, the Swenson Ford Thunderbolt finally had its manual transmission and was readied for action. Al performed his proven thin-seat valve job and installed his rocker arm-shaft end supports. With the High Riser's higher-lift camshaft the lash settings were tightened, but not nearly as much as in previous years. Because the class rules required cheater slicks no wider than 7 inches, the cam was retarded 4 degrees to take advantage of mid- and high-range power. In a further effort to tailor the powerband, the transmission was fitted with a lower-ratio first gear while either 4.89 or 5.12 rear-end gears took advantage of the High Riser engine's willingness to rev to 7,000 rpm.

Al determined the average best mean torque to be at 38 degrees of total timing. Once he arrived at this average, Al then worked to dynamically change the timing, particularly in high gear. This is where his proficiency with the Sun distributor machine came in handy. He discovered that dropping 3 degrees of total timing in high gear, where a heavily loaded engine requires less advance, resulted in an improvement of nearly a tenth of a second in elapsed time and two mph in trap speed. Al added a switch that would cut out one of the distributor's two sets of ignition points in high gear in order to accomplish the change in dynamic timing. Hard-learned lessons regarding carburetor jetting in relation to altitude, temperature, and humidity had taught Al to rely more upon the trap speed indicated on his time cards, rather than upon visual examination of the spark plugs, or on elapsed time, which was affected by traction.

Hard work soon paid dividends as Al began to find regional success, taking home class and event wins, and once again gaining the coveted title of *Eastern Drag News* Top Dog. On June 21, 1964, *Eastern Drag News* reported, "Ford Leads the Parade at Atco's $1,000 S/S Meet." The action was described as follows:

> "Favorite on the betting boards was hometown hero Bob Harrop and Ramcharger Dodge. Not to be counted out of the picture was Al Swenson's Ford entry driven by Al Joniec. During the first round of eliminations it was these same two cars that were doing the eliminating. Out of the many Ford and Dodge entries, Joniec and Harrop were the last to hit the home stretch. With $400 riding on the nose of the winner, both cars were hot out of the hole heading for pay dirt. It was strictly driver's skill that won this race as Joniec went through the gears a little faster and pushed the Ford Thunderbolt out in front for the lead and the old green stuff. The Ford fans went wild as Joniec hit the Chrondeks at 11.73 and 119.04. Harrop followed it up with an 11.85 and 117.60."

Along with the Thunderbolt came a more professional-appearing and safer means of transporting the race car to and from the track. After experiencing a number of harrowing near misses while flat-towing the 1962 and '63 cars, Al was presented with a used Ford flatbed truck from the Swenson Ford used vehicle inventory. But it didn't turn out to be the big step up from flat-towing behind a station wagon that it appeared to be. The truck was a flatbed and not a rollback style, which meant that long, ungainly ramps, along with a downhill slope, were required to load the car. On top of that the truck had no room for storage, and lacked power steering, power brakes, air conditioning, or any other options, making for a hot, bumpy ride. The oversized ramps resulted in Al smashing his thumb getting them out from under the chassis where they were stored. Since then he has retained the thumbnail he shed that day in his tool box, as a reminder to always be careful what you wish for.

THE 1964 NHRA NATIONALS— A BRIDESMAID ONCE AGAIN

Attendance at the 1963 Nationals, held in Indianapolis on Labor Day weekend, had taught Al a lesson about how cars perform in hot, humid weather.

He vowed to be ready for these conditions at the 1964 event. The Super Stock class had more than 100 entries, most of which were Thunderbolts, with a few Chrysler products, such as Dave Strickler's Hemi Dodge, thrown in. Al again leaned out his carburetor jetting to compensate for the humid conditions and was pleased to find during time trials that his car was among the quickest in competition. But, as is the case with any true competitor, he wanted to be THE quickest. Most—if not all—of the entries in the Super Stock class were using M&H Racemaster cheater slicks (NHRA rules for the class required that rear tires be no more than 7 inches wide and have two grooves covering the circumference of the rear tires. Thus the term cheater slick, as the face of the tire was not a full slick). Bob Harrop, who was competing in the Super Stock Automatic class with his hot Dodge, offered Al the use of a new pair of Goodyear tires. Because the Ford and Mopar rims were interchangeable, it was

Even though he was once again the bridesmaid at the 1964 NHRA Nationals because of a clutch problem that struck in the semifinal round of competition, Al still managed a smile in this photo.

Here's Al in action at Atco Dragway in the Al Swenson Ford Thunderbolt. In the other lane is Lon Razer in the former Al Swenson '63 lightweight Galaxie that Al campaigned successfully the previous year.

Getting the Thunderbolt off the starting line on 7-inch-wide cheater slicks always proved to be a challenge with nearly 500 horsepower under foot. The slight haze of smoke off the rear tires in this photo shows that Al had mastered that delicate dance.

an easy swap. The tires were taller than the M&H's and were mounted on a wider rim. To compare their characteristics Al took the T-Bolt to a remote area of the pits to perform some test burnouts. A study of the patterns left on the asphalt by the tires during burnouts showed they had inadequate contact in the middle, so Al increased the air pressure in the tires to 20 psi. The results were encouraging as the car immediately recorded a quicker elapsed time than it had previously. However it was also noted that the car gained no mph due to the Goodyear tires being taller. Al installed 5.12 ratio rear-end gearing prior to the next day's class eliminations.

The following day competitors were informed that because the number of entries in the Super Stock class, the winners of each round of competition would have to return to the staging lanes immediately, offering little cool down or maintenance time. In the high heat and humidity the Swenson Ford T-Bolt rose to the occasion, running elapsed times in the mid- to high 11.60s. After winning the first round easily, Al opted to raise the tire pressure once with each round, eventually pumping them up to an unheard-of 40 psi.

Raising the tire pressures and keeping the intake manifold cooled with copious amounts of ice did the trick. In the meantime a Goodyear engineer, having heard how well Al was doing with the tires, came over to determine the reason. Upon checking the rear tire pressures with his gauge the engineer was astounded to find them at 42 psi.

After five rounds of competition, Al found himself among the final three cars left in Super Stock, alongside Butch Leal (Mickey Thompson T-Bolt) and Len Richter (Bob Ford T-Bolt). NHRA officials decided that the three competitors would flip a coin to determine who got the bye run (solo), while the other two ran all-out against one another. When Butch Leal won the coin toss Al took exception, because Leal had already had two bye runs in eliminations. An argument ensued and Al instructed Leal to stay put, as he would be back for him after winning this round.

Against Len Richter, Al performed his burnout routine to heat the rear tires and then staged the car in the starting beams. As he began to bring up the rpm to launch he felt the car move slightly against the line lock (a button-operated momentary switch/hydraulic device that locked the car's front brakes to keep it from rolling out of the starting beams). Pushing the clutch pedal to the floor did nothing to slow the car's forward progress, and it slid forward, breaking the starting beams and causing a foul start. Crestfallen, Al limped his car back to the pit area, where it was determined that the clutch linkage had bent due to stress and high heat, costing him yet another chance at a National event win. Butch Leal would go on to win the Super Stock class title at the event, proving the old adage that it is often better to be lucky than good.

"I was in shock and somewhat depressed as another one got away," Al said. "All I got was the satisfaction that I really had the fastest Ford yet again, with literarily all of them there (including the Ford team cars) at one place and time. All the Fords were invited to work out of Bob Ford Dealership service garage in downtown Indy, and it was a great place to work and BS with the other guys. I finally had a face-to-face meeting with most of the Ford racing management and Ford racing team guys. It was great, as I felt like one of the guys. However, I got no sign from anyone that I was going to be anywhere but on the outside, looking in."

Al Joniec would close out his 1964 season with additional regional wins, coming as late as November when he took home the Super Stock class win at Vargo Dragway. His reputation as one of the best Ford racers in the nation was growing—he never lost to another Ford after converting to a four-speed transmission. After season's end the Swenson Ford Thunderbolt was sold. It disappeared from history and remains unaccounted for to this day.

CHAPTER FOUR

AL RECEIVES THE CALL UP

WITH THE WILDLY SUCCESSFUL introduction of the Mustang in the spring of 1964, Ford decided to make the Mustang the carrier of its performance banner for 1965. Ford's intentions for the Mustang were no secret to Drag Racing Coordinator Dick Brannan, but even he was unable to secure a pre- or early production Mustang for drag race development. Because the first-generation Mustang shared its platform with the Falcon, Brannan decided to have Dearborn Steel Tubing build him an A/FX-class Falcon, powered by the 427 High Riser engine, to get a jump on the competition for the upcoming season. Ford ultimately tapped Dearborn Steel Tubing to construct a 1965 Mustang fastback, powered by the SOHC 427 engine that had been rejected as viable for NASCAR stock car competition, as a "mule" for a proposed program to supply select members of the Ford Drag Council with Mustangs for competition in the A/FX class. This car was further developed by Brannan and his team.

The contract to build the Mustang drag cars was ultimately awarded to Holman-Moody (A Ford performance sub-contractor best known for preparing stock cars). Holman-Moody was to build 10 additional Mustangs, to be delivered to Drag Council members prior to the running of the 1965 NHRA Winternationals in Pomona, California. Al Joniec was not yet aware he had been chosen as one of the Drag Council members for 1965, but as Dick Brannan himself told this author, "Don't you forget to tell them that I chose Al and he was an ORIGINAL member of the team."

Meanwhile back in Philadelphia, Al was totally unaware of the goings-on in Dearborn, or his pending selection as a member of the Ford Drag Council, as can be seen in his personal recollections.

Right after Indy I got a call from Bill Jenkins. He informed me that he got a deal with Chrysler. He asked me to drive the new car because he had split with Dave Strickler and they were no longer a team. I said yes, as I was thrilled to finally have some kind of direct factory deal and especially with Jenkins, even though I would have preferred a Ford (most of the top guys had changed to different manufacturers).

He called me shortly after that to tell me the car was arriving and to come on down and see it. It was supposed to be a 4-speed but came through as an automatic. So Bill said, "Hell, I can drive it and I really didn't need you anymore," but I did help him prepare the car for the 1965 Winternationals. I had never visited California and was looking forward to going anyway.

Because the car was so light he had to add weight, especially in the rear. One of my high-tech jobs was to remove the taillight housings and add weight. I put a pipe around the light socket (to protect the socket) as we poured molten lead into the tail light housing until it was about halfway full. I also put long screws in the light housing for the lead to grab onto and painted the leaded area with sliver paint. After the modifications, the taillights still worked as designed and you would never know from looking at them that there was at least 20 pounds of lead in each of them. Jerry Stahl made him some great headers while Bill took the heads off to do his thing with the valves.

When the time came close to leaving for California and the Winternationals, Bill told me he didn't have the space (he towed the car with a 1956 Chevy wagon) or resources to bring me along. I understood why, but was now totally bummed as it was like an emotional rollercoaster for me. I felt like a rubber glove after a proctology exam.

As Christmas 1964 approached, with no word from Ford and the deal with Jenkins dissolved, Al found himself second-guessing his future in racing. Then one day the phone rang, and on the other end was none other than Charles Gray, who just so happened to be "the Man" when it came to Ford drag racing at the time. Gray informed Al that he had been selected as one of the Ford Drag Council members to receive and campaign a new A/FX Mustang for 1965. Al was instructed to pick up his new mount at the shops of Holman-Moody in Charlotte, North Carolina. On December 16, 1964, Ford Motor Company issued a manufacturer's statement of origin to a motor vehicle for one 1965 Ford, model 09, two door fastback VIN 5F09K380235. This car was shipped to Holman-Moody for race preparation and would become the Swenson Ford-sponsored A/FX Mustang, which was sold to Al Joniec for one dollar.

Nineteen sixty-five turned out to be another educational year for me as the early 60's Super Stock years started to evolve into the lawless Funny Car years. Prior to then there were match races, but they usually had unwritten rules between the racers to keep it relatively stock. This started to give way to "Run What You Brung" in the South, even before 1965. Eventually this style of racing crept everywhere. So by 1965, "shit was happening," as I started to see acid-dipped bodies and altered wheelbase cars with big engines and injectors sticking out of the hood— some running on nitromethane. There simply were no rules.

Then I got the call to go to Holman-Moody and pick up my new 1965 AF/X Mustang. They apologized that there were not enough SOHC 427 engines yet available; I would be getting a

These snapshots are the first known photos of Al Joniec's 1965 A/FX Mustang, taken when he returned from a trip to Holman-Moody in North Carolina to pick up the car.

standard Wedge. Hell, I was happy and didn't care what kind of engine it had at that time. I borrowed a decent trailer to go get it, as there would have been no way to put it on last year's flatbed truck (it was sold while I was gone). Here I was, a kid from Philly driving to Charlotte, the real South and a place I never been, except to drive through. I was now looking at White-only bathrooms and Colored-only water fountains—what a culture shock!

But Holman-Moody was an amazing place. I had never seen a facility like that before. Here was this enormous garage separated by a chain-link fence right down the middle. One side was all of the FoMoCo stuff. The other side was H-M NASCAR team cars with parts everywhere. The completed engine pallets alone were overwhelming to me. They were stacked stories high; I couldn't fathom how many were there. They had GT40s, both complete and partial shells in the queue, all being worked on by a beehive of guys. In the background was the constant noise and hum of many dynos being run. I was in a different world. I felt like a little boy, taking it all in.

I don't remember an office (I'm sure there was one) and there were no formalities. I just kind of walked into the first open garage and asked the first guy I saw, "Who do I see about picking up my car?" A guy named Red (he had red hair) took me to the garage stall were my car was being worked on. It was far from being complete, as they were just starting on the headers. Oh yes, the headers! They were a work of art—they were being made out of incredibly thin, .030" wall stainless 2" tubing. I was so appreciative of the skill level these guys had, especially remembering when I did my crude headers in 1963. Beautiful craftsmanship and ultra-light weight—I was just thrilled with them.

While we were working on the headers there was a rain storm and the gutter pipes were overflowing. One of the good ol' boys remarked in his finest Southern accent, never missing a beat from his work, "Sounds like a cow pissing on a flat rock." I laughed so hard I walked away in tears!

I met both Ralph Moody & John Holman, Paul Norris (H&M's drag guy), Fred Lorenzen, Dan Gurney, and so many other NASCAR guys that I can't remember now. It was unbelievable to me, as I thought it couldn't get any better than this. I felt like I died and gone to the great garage in the sky.

The car looked like a regular Mustang. The only big real work was in cutting out the spring towers, and adding the rear traction bars and roll bars. The front replacement springs were ingenious; they were simple, flat springs that were firmly mounted to the frame and the lower A-arm of the suspension. This made them act as a flat torsion bar. They were brilliantly simple, lightweight, and they did the job.

Once I brought the car home I was ready to go to work. I had it painted mother of pearl white by Bob Cummons (Bob was a kid who hung out and helped at Swenson when we working on the race car at night, who had opened a paint and body shop). I named it *Pegasus*, after the flying horse of Greek Mythology. The car was lettered by the sign painter that did Swenson Ford's window promotions. After this, only Ralph Hart (who I met and became friends with in the Bob's Big Boy years) did all the art work on my cars, and only Bob painted them.

SOMETIMES THE TRIP IS AS MUCH FUN AS THE DESTINATION

"When I got home from picking up the Mustang from Holman-Moody, I found that Ford had sent me an invitation to the 1965 Ford Motorsports Banquet in Dearborn. I didn't know what to expect, but I was more than looking forward to it.

Though I had flown in small planes I was never on a commercial jet before. I thought it would be another adventure, and boy it was. I was standing in line at the Philly airport in my new suit and six-button Benny overcoat (it was near Christmas and it was really cold).

Al had finally made the big leagues of drag racing as can be seen in this photo of him with his Swenson Ford 1965 A/FX Mustang and top-of-the-line ramp truck transporter.

I couldn't help notice the other people in line were pointing and talking about someone. So I asked the people ahead of me what was going on, and they pointed to a girl who was slumped over sleeping while sitting on her mink coat. They said that she was Diana Ross of the Supremes. Holy shit! I said to myself; Diana Ross was easily the hottest entertainer in the country at that time, with around seven number-one records in a row.

The line of passengers started to move. Everyone was fixated on her as they walked past, nearly touching her, but not me. I was feeling good so I leaned over and grabbed her by the arm and said, "Come on, honey, you're going to miss your plane. Let me help you."

She woke up a little and said "Oh, thank you." As she stood up I grabbed her mink coat and threw it over my arm while I held her with the other. We started walking toward the plane and she started to wake up some more. She realized I was really helping her, and she became more appreciative and held onto my arm tighter as she started to talk. She told me she ran away from her group and had a much-needed date with one of the Philadelphia Eagles football players.

By now we were approaching the plane as several stewardesses stood on the ramp entrance looking for her (apparently they had received word that Ms. Ross was on the flight). I could see that the stewardesses were excited—I assumed they thought I was either her boyfriend or manager, as they sat us both in the best first-class seats.

They never even asked me for my ticket (my seat assignment was in the cheap seats in the back).

So here I was, never having flown on a jet before, but now in a 707, going to a dream event, sitting in first class, eating fresh shrimp and drinking champagne. Even better, I was talking to one of, if not *the* most popular entertainer in the world at the time, and listening to her incredible life story.

Though she was thin she was so very well groomed. With her hair and makeup, high-end clothing and diamond jewelry—the whole nine yards—she was impressive. She also never stopped talking about how she grew up singing on the front steps of her ghetto house when she was a kid. She explained that she had never taken a singing lesson, and just started to sing with other girls in the neighborhood. Well, after all, she was from Motown.

She told me how the other girls in the Supremes hated her and that they would be breaking up after their contract ended in two years (she was telling me the truth). She also showed me her schedule, which went on for years. I just couldn't believe it—it looked to me like she was set to appear at least 2 to 3 times a week everywhere around the world. As a matter of fact, she would be on the Ed Sullivan show that night. The stewardesses kept coming up to me (I was in the aisle seat) and asking me if they could get her autograph. It was all surreal.

When we finally landed, she wrote on a piece of paper, "To my traveling Buddy Al, Diana Ross." There were a lot of people waiting for her when we arrived and they looked pissed, so I started to get concerned. They ran right up and grabbed her before we could formally say goodbye, but she did turn around and wave to me...

Upon arriving in Detroit I rented a car and was trying to find the hotel. While waiting at a stop sign, I felt someone pushing my car from behind (the cars had real bumpers back then). I stood on the brakes as the other car tried to push me into traffic and looked back to see Bill Lawton and John Healey in the other car, laughing their asses off at me. We were like kids the next day as we raced and played bumper cars with our rent-a-racers. I also learned how to blow the muffler off the rental car. You just had to be going fast, put it in low gear for maximum deceleration fuel enrichment (this works only with carburetors), shut off the key to let all the raw fuel accumulate in the muffler, and then turn the key on—BANG! It was like setting off a hand grenade. If you did it enough times the muffler would blow apart. The car rental people must have loved it when the racers were in town.

The Ford banquet was huge and incredible. I had never experienced anything like it before. I was seated at a table with Don Prudhomme, Ed Pink, Lou Meyer of Meyer-Drake Offenhauser engines, Cale Yarborough, and someone else I can't remember. They had an open bar and this huge, five-foot bowl made out of real ice. They kept bringing in freshly cooked, steaming hot, giant shrimp with several buckets of cocktail sauce next to it, and I was going at them big time.

They then asked us to take our seats. The lights went out in this giant place the size of a football field. Out of each of the four corners of the room came a big, flaming table being pushed by chefs and followed by an army of servers. On each table was a huge half a cow, flaming enough to light up the room, and they merged in the center of the room. The lights came up, the flames were blown out, and the chefs started to carve off the meat as their assistants served the guests. The whole thing was incredible and was one of the best meals I had ever had up to that time. The announcer called all the drivers up to the stage. Lee Iacocca presented me with a gold pen with my name engraved on it that I still treasure to this day. Lou Meyer really gave me a good education regarding a problem I

Al's A/FX Mustang is fully decked out in Swenson Ford livery and bearing the name *Pegasus*. Wider slicks, improved rubber compounds, and of course BX-10 improved traction greatly over the previous year. The 1965 Super Stock Magazine Nationals, held at York U.S. 30 Dragway in Pennsylvania, has been called the Woodstock of Drag Racing. All the biggest names in S/S and F/X racing from across the country were entered. Al's Mustang was still powered by the 427 Wedge engine at this point, which allowed him to compete in the 3000-lb. gas class. Mustangs equipped with the SOHC 427 ran in the 3,200-lb. class.

had been having with rod bearings. He set me on the path to approaching everything from an engineer's point of view from then on.

I was pumped after that banquet and couldn't wait to get started on the Mustang. I went after the engine first by taking off the heads and looking at what we had to work with. The car was set up to accommodate a much wider Cammer engine, so the Wedge had plenty of room to work around it. I just loved the easy-to-work-on H-M custom headers, as they came out individually instead of in one piece. The car had 10.5-inch slicks, which were way better than the 7-inch tires we had to run in the previous year. But even with the 10.5" tires, traction was not the best (unlike today's 10.5 tires). There was too much weight on the front end because of the standard engine mounting position and short wheelbase. The car did get into the 10-second range right away though, as the traction was much better with the Wedge than the much heavier Cammer.

I started testing and winning races, but there were not a lot of AF/X cars around to race locally at the time. Ultimately the performance difference between the Wedge and the Cammer was very little in elapsed time, but the Cammer was about 4 mph faster due to their additional power. The first Cammers had steel heads, which were 80 pounds heavier than the aluminum ones. Even with the aluminum heads, the Cammers were still about 50 lbs heavier than the Wedge. That was why the ETs were not that much different. Plus, all this weight was over the front wheels. Traction was so bad with the heavier steel-head Cammer that I started looking for an alternative to the rosin we used to enhance traction in match races.

Tire technology had not kept pace with the amount of power that drag race engines were capable of producing by the mid-1960s. As a result many tracks allowed the use of rosin to aid traction during match races. Also known as "gold dust" because of its color, the rosin powder would be spread on the drag strip surface at the width of the rear tires, and then a broom was used to embed it into the asphalt. The cars would then perform burnouts across the starting line through the rosin. As it got hot from the friction, traction would increase dramatically. Watching the spreading of the rosin and subsequent burnouts was very popular with fans, this writer included.

I came up with a way to liquefy the rosin. I used a solvent to keep the rosin in a liquid solution so that it could be painted on the tire and then dry quickly. It worked. The problem was, you picked up a lot of dirt driving to the line, so you had to do a dry burnout over the line. Once you did that you had a clean tire and rosin both on the tire and on the track (as long as you backed into your own tire marks). It was good for a solid two-tenths ET. This creation was later marketed under the BX-10 name.

One day Ford asked me if would like to try out a new trick camshaft with .650" lift. They warned me that, due to the additional lift, the valves might hit the pistons, and I would most likely have to cut them. I said, "Hell yeah," anything for the cause. Sure enough, when I put the cam in and checked there was only about .030" actual valve-to-piston clearance. With that clearance you could start and

As part of his association with friend Jim Rahm's company MAI, Al was head of the company's chemical division. He developed a liquid traction compound marketed as BX-10. The BX stood for Brand X and the 10 stood for the tenth of a second improvement in elapsed time the customer could expect to gain after applying the product to their rear tires.

run the engine with no problems, but you couldn't race it because of valve toss (not valve float).

There was a big NHRA regional race at Cecil County Dragway coming up that I really wanted to attend. To solve the valve-clearance problem I had to choose between the long way of taking out the engine and cutting the pistons, or I had to try to make up a trick cutter and cut the pistons with the engine in the car. I went with in-car method. I was working with a Mr. Schmitt, a retired German machinist from the old country. We took some old valves, brazed two small pieces of tool steel on the faces of the valves, and then machined them level. He then removed the valve material between the cutters so he could then sharpen them on the cutting edge—think of a fly cutter on a valve stem. It worked great. I also had an adjustable, precision collar made for the valve stem. When I put the piston at TDC and lowered the cutter onto the piston, I could adjust the collar stop to the depth I wanted to cut (.100" in this case). I had to do each valve, one at a time, by bolting on the head using just two bolts, After adjusting the fly cutter, I would chuck the valve stem in a drill and start cutting, pressing down until it hit the stop. We covered and taped off everything and put lots of oil on the cylinder walls to catch the aluminum chips from the cutter. We simply moved the piston from TDC to BDC, wiped and applied more oil until there were no more chips to catch, and in a couple of hours we were done. Boy, the engine sounded stronger too, and all in time for the big NHRA Cecil County race.

The Ford race team manager Charlie Gray was there. I don't remember who else I raced, but I knew that Tasca Ford had the car to beat; he had a Cammer and I had a Wedge. I picked up around two tenths with the new trick cam (from the 10.90s to the 10.70s), so I was happy. I'm not sure how many cars I beat, but it was between Tasca (who was running in 10.60s) and me for the trophy. I put a big-time hole shot on his ass and held him off and beat him. Boy was Old Man Tasca pissed when I won.

MOTOR TUNING is all in a day's work for Joniec as he makes last minute adjustments

In this scan from a period car magazine, Al can be seen working on the SOHC 427 engine in his Mustang between rounds at a race during the 1965 season.

After the race the Ford guys asked me what I did to run so well. They said that Bill Ireland and Clester Andrews (who also had 427 Wedge-powered AF/X Mustangs like me) both got the same trick cam but they bent their valves. I may have gotten the first available SOHC because of it. At the now-very-famous York Super Stock Nationals (the Woodstock of drag racing) there were 86 Factory Experimental type cars competing in various weight classes. I won the 3,000 lb class with the Wedge engine. The Ford team cars with Cammer engines were in the 3,200 lb class because of their extra weight. Another one I remember was a 1/8th mile winner-take-all race in Manassas, Virginia. Once again the race had all kinds of crazy funny cars and me with a carbureted AF/X machine. Ronnie Sox had two cars in the race. I was winning round after round when I came up against Ronnie's # 2 car (the slower of the two), and I beat him outright. I was real happy to have won, but as I came back on the return road I could hardly see from all the smoke and dirt flying around. Ronnie's car was on its side. As I got closer to the car the door flung open and there was a pair of feet dangling out. I got out of my car and ran over to help pull him up and out. He jumped off the side

At the 1965 NHRA Nationals, Al appeared to be well on his way to victory in the A/FX class until an uncharacteristic missed shift against eventual winner Les Ritchey's similar Mustang dashed his hopes once again. Here Al defeats Gas Ronda in the early going.

of the car and looked at it and shook his head. He was physically OK. It took some time to clear the track and, don't you know, it was Ronnie (he was a hell of a man) and me for the final run, but this time he was in his #1 car. I got him out of the hole (BX-10 really worked) and almost won, but those extreme funny cars were just too fast against my legal AF/X.

I got another call from Ford. This time they said that Paul Norris of H-M had crashed badly (because he was in the South, he was running a "run what you brung" car by that time). His car was beyond repair, so I would be getting his Cammer engine, complete with injectors. The engine arrived strapped down to a covered wooden pallet. From the looks of the engine, Paul had tumbled into the dirt many times. Grass and dirt were all over the engine and inside the broken injector stacks. It was a real mess.

I had to take the heads off to make sure that the engine didn't get hurt by ingesting some dirt. This was quite an education, as I really had never seen the inside of a Cammer before. Luckily the engine was OK and I just had to clean it up, but I couldn't believe the weight of the steel heads. I had to find the rest of the parts I needed, such as a dual-carb intake manifold, oil pan, and headers. Once I got all the parts I made the swap with no problem.

When I took it to the track I could feel the additional weight on the front end, as it hurt the traction. With the steel heads the car ran only a little better than the Wedge, mostly on the far end due to the additional power. I took it to the new NHRA Summer Nationals at Bristol, Tennessee. I almost won the final run when the engine mysteriously lost some power. I didn't know that a head gasket was starting to leak earlier. It completely let go on the final run and overheated the engine. Charlie Gray saw that I was working with old worn out junk, so he said he would send me a new engine and some new aluminum heads soon afterward. I started closing down the valve lash like I did with the Wedge, but on the early Cammers there were no adjustable rockers. You had to use lash caps of different thicknesses to adjust the lash. I asked for a selection of different lash caps and I got a lot of them, but you never have enough when there are 16 different size requirements, and I wanted to do a lot of small changes. I came up with a fix by having a 3/8" punch and die made by my machinist, Mr. Schmitt. Using this punch, I could make my own valve shims. I would buy various feeler gauges and simply punch out the sizes of shims I needed and put them under the valve caps. I now had tons of shims instead of tons of lash caps, and it gave me much more flexibility.

Sure as hell, the engine liked less valve lash, just like the Wedge did. Because the stock cams only had .550" lift, any additional lift was great for the engine. Eventually, after many, many complaints, Ford finally made adjustable rockers.

I then started with cam timing changes. This was easy, with adjustable cam timing gears conveniently located right under the cam covers. I advanced and retarded constantly in search of the best combination. Again, the retarded position worked for me. Even though I tried doing it, I didn't believe in doing one side differently than the other due to the extra chain length from one side to the other. If the chain moved it would just retard the cam timing even more, and that was good as far as I was concerned.

There are a number of tools you can use to tune your specific engine combination. Usually, if you retard the cam and the engine responds with more horsepower, it means the cam is not big enough. Typically when you retard the cam you lose some mid-range power in exchange for top-end power. Conversely, when you advance the cam you increase low- to mid-range power and lose some top-end power. If you advance the cam and the engine responds with more horsepower, it usually means you have too much cam. Less valve lash adds both duration and lift, while increasing valve lash reduces duration and lowers lift. By changing your rocker arm ratio, you can also change both duration and valve lift (the secret Achilles heel of the Cammer was its 1.3 rocker arm ratio). If you get results from any of these adjustments but can't go any further, it then justifies buying another cam to take you to the next level.

The cams in the SOHC 427 were too conservative, as the original intent for the engine was NASCAR competition. Even though I could rev it higher, the power rolled over at about 7,800 rpm, about the same as I could get from my Wedge engine. For drag racing applications you needed to go even higher in the rpm range.

When the aluminum heads finally arrived life was good again, as around 80 pounds came off the front of the car. I could now add the weight I needed in the rear for traction. Boy, you could feel it driving the car—it immediately went faster. I was now finally getting good traction to go with the additional power of the Cammer. When I found out Ford was working with Crane Cams I enthusiastically said I wanted/needed to try them, as I was starting to get ready for the NHRA Nationals again and needed something better than the conservative standard cam. Crane came up with a .600" lift cam and everyone on the Ford team got them just before the Nationals. I put mine in with no problems and went to the local track to test. You could feel the difference right away; the car really went faster as the engine powered all the way to 8,500 rpm. I put in a 5.12 rear to take advantage of the additional rpm and the car was never faster, now running in the mid 10's.

However, the additional rpm caused a new and bigger problem to surface. The rod bearings were spinning as I was losing oil pressure (the main bearings looked fine, though). Apparently the additional g-loads from the higher rpm were stretching the rod caps at the parting line, pinching off the oil and taking out the bearing. I took off the pan and sure enough, several bearings were in various states of failure. Some were completely spun and others had their locking tab bent over.

This was my new engine and I hadn't done anything to the bottom end. Here it was, the week of the Nationals. I didn't know what to do other

Al and his friend Jim Rahm at the NHRA finals in Tulsa, Okla., in 1965. After yet another loss caused by a missed shift, Al and Bud Rubino discovered the cause of the problem and devised a fix.

than to replace all the rod bearings and go. I did take as many extra rod bearings as I could get, at least 16 as I remember. I kept some old but OK-to-use ones as well.

We got to Indy and were pleased to find out that it was not as hot as the previous year. I was looking forward to the race as I was now in one of the Ford team cars; I believe all the team cars were there. There were not as many cars running in A/FX. Everyone had expected the Ford team to win so we basically ended up racing against each other. I did lean out my jetting some before I got there and was ready to race, complete with a new set of Goodyear tires we got at the track.

Off the trailer I was the fastest A/FX car from the start this time. I was now shifting around 8,500 rpm and when I flat-shifted it (meaning, I didn't take my foot off the gas between shifts), the engine over-revved to 9,000 to 9,200 rpm as indicated by the tattletale needle on the tach. Like I said earlier, to win you have to be mean to your machine.

After every run I pulled the pan to check the rod bearings. Sure enough, there were always at least 2 to 4 (sometimes more) rod bearings that had started to move, as the locking tabs were taken off and needed replacement. I felt that as long as I was catching the bearings before they actually spun and not ruining the crank and rods, I could keep doing this. After the last time trial, we went back to the motel to work on the rod bearings and get ready for the next day's eliminations. I clearly remember Bill Lawton looking over the railing of the motel room that we all stayed at. He was laughing his ass off as he watched us working most of the night replacing every rod bearing and prepping the car for the eliminations (I believe we replaced more than 32 rod bearings over two days). He was drinking beer, pointing his finger and heckling us, genuinely having a good time. I miss his incredible smile and sense of humor.

"I had to borrow every rod bearing from the Tasca team (John Healey became a good friend, and we still are friends today) and anyone else I could in order to stay in the race. I'm not sure how it happened, but John and I raced each other for the first run of eliminations and I won fair and square. My guys told me that Bob Tasca threw their ice chest in the air in frustration and yelled, "That damn Joniec!" but he did come over to congratulate me later. The next round was against Gas Ronda, and I won that one as well. By now we had become experts at replacing rod bearings. I then faced Les Ritchey. Even though I had been running in the low 10.60s and Les was running in the high 10.60s, a new and mysterious problem appeared in our race, as I missed third gear and lost. After that Les went on the win the event. This was the fifth time I lost at the Nationals, as yet another one got away.

A short time later Ford engineering said they identified a heat treatment problem with the rods, which gave me and others rod bearing problems.

"There was one more race at the end of the season at Tulsa, Oklahoma, called the World Finals. I had been checking and replacing clutch and transmission parts in an attempt to find and correct the shifting problem I had at Indy, though I found nothing. Because I didn't find anything

After taking delivery of the SOHC 427 engine from the wrecked Holman-Moody A/FX Mustang of Paul Norris, Al obtained the official Ford service manual to assist with the learning curve for the sophisticated new racing engine.

and I wasn't sure what to do, I went to Tulsa. I was the fastest, and most of the Ford team cars were there again. This was a top stock eliminator handicap start race, so anyone could win, even with lower, stock class cars. I lost due to a missed shift once again, and I'm not sure who I lost to or who even won the race, as I tried to forget bad experiences like this.

I was now not only disgusted with it all, but now I was getting depressed. I was about to load the car to head home when I asked Bud to push the clutch pedal while I watched the linkage. When he pushed slow everything looked OK. But when I asked him to push it as fast as he could, like I would when I shifted it during a race, the entire firewall on the driver side buckled and/or tin canned around two inches. I couldn't believe what I was looking at, but at least I finally found the problem. Apparently the firewall metal had fatigued and started to move. I had installed a high spring-pressure clutch and I guess that was the straw that broke the camel's back. To fix it I simply put in a support brace.

BLOWING OFF STEAM ALMOST SENDS AL AND JIM TO DAVY JONES'S LOCKER

After the race in Tulsa, Jim and I went to the West Coast. Jim planned to go to some speed shops to sell BX-10 and one of his stops was Vic Hubbert's, who was into race boats. Jim and Vic were negotiating as I was falling in love looking at this absolutely trick 427 Ford-powered ski boat. I found out later that Vic wanted to be the West Coast distributor for BX-10 and Jim cut a deal with him to take the entire truck load for the boat to seal the deal.

The next thing I knew we were hooking up a trailer hitch. We continued up the coast to see Jim's brother in Spokane, Washington, and eventually we towed the boat 3,500 miles before we got home. Every time we stopped for gas or food a huge crowd always gathered around the boat, so we knew we had something special.

After we got back home it was fall and the weather was cold. But we and our friends hoped to be able to take the boat out on the water when the weather was right. One Indian summer day (68 degrees), Jim and I couldn't keep our pants on any longer, so we took it to the launch ramp in the Delaware River. Even though years before I was on a destroyer in the Navy, neither of us had ever handled anything other than a row boat before. We didn't know a thing about nautical things, and here we were going to drive a "Widow Maker" water rocket.

I did remember that Vic told us to make sure we kept the bow high and never let it get below the water line, as it could drive the boat under. The weather was absolutely perfect with no wind, and the river was as calm as I ever saw it. When we launched the boat we thought we had died and gone to heaven as we idled out of the ramp area to warm up the engine.

We really didn't know what to do, but we knew we wanted to experience the power and speed. Going

fast on the water in an open cockpit a foot of so off the water was a brand new experience for the both of us. I was driving and I just floored it and was kind of disappointed at how slow it accelerated from idle. The prop just cavitated as the boat struggled to get up on plane. It was not like today's flat-bottom speed boats that take right off. It was designed to ride only on the two sponsons on the hull and half of the propeller. Once we were up on plane, boy did that sucker take off. It made an absolutely huge rooster tail of water. Only about half the prop was in the water, and it threw spray seemingly hundreds of feet in the air.

Once on plane it accelerated like my race car, it was so fast. When we got up over 80 mph or so, the tears from my eyes from the wind went straight back. We didn't have any googles and no windshield. It was a wild ride on a perfect day and we were just having a great time flying on the water.

Because the river was so smooth that day, cruising around 50 mph felt like no big deal. We quickly went too far, so we decided to turn around to go back. There was hardly any boat traffic that day, but when I turned back I could see a big tugboat coming toward us about a mile or so away. At that time I believe Philly was the fourth-largest port in the US, and there were a lot of big boats that needed big tugs to help navigate the river.

For those who didn't know, tugboats are very heavy and deep in the water and easily make the most dangerous wake. Only problem was that I found out these tugboat facts later, not before. So we were going about 50 mph and as I got closer to the tug I could see that the wake was really a swell of rolling walls of water about 6 feet high. I really started to pucker as I saw the wake of the tug. We both looked at each other in despair, wondering what to do. I still remember the tug boat crew coming out of cabin to watch the disaster that was about to take place. In the back of my mind, Vic was reminding us not to let the bow go under the water line, so I jammed the throttle to full as we approached the first wall of water.

We hit the first wave and it launched us in the air about 10 to 15 feet—but we never hit another wave as we flew over the rest of the entire wake. It was surreal as time seemed to slow way down. I clung to the steering wheel with all my might and saw the tugboat crew looking at us in awe, elbowing each and other pointing at us as we flew through the air. I thought for sure we were going to get hurt, but we landed completely flat on the water. A huge, splashing wall of water engulfed the entire boat and us.

The engine stalled from the water and we were both soaking wet from head to toe. The tugboat crew were laughing their asses off, as they had a front row seat to it all.

As the tugboat drove into the sunset we both looked at each other. Without saying a word, we knew we had just dodged another bullet. I bumped the starter motor over to make sure the engine didn't hydraulic itself to death. It finally started and we slowly idled back—and I mean slowly—to the launching ramp before riding home, still soaking wet and silently reflecting on the "Holy Shit" day.

CHAPTER FIVE

HOLY HEMI, IT'S THE *BATCAR*!

WHEN I RETURNED HOME after Tulsa and came back to reality, I was again genuinely enthusiastic and looking forward to the next year. The Ford banquet really did it for me again, also. As I started to assess the 1965 season, one of my disappointments was that there were so many other cars called *Pegasus* out there, and I wanted a change. Right after I had that thought, I was eating at a bar watching TV when the new Batman television series came on, and I instantly fell in love.

The Batman series featured the Batmobile, a campy take-off of the Lincoln Futura concept car from the 1950s, which had been sold to California customizer George Barris. The Batmobile was painted black and accented with red scalloping, which immediately struck a chord with Al, as it was similar to the scalloped paint scheme that he had loved on Lou Tomaselli's 1958 Ford some years before. Al had Bob Cummons strip the Mustang and paint it black, and he contracted Ralph Hart to devise a red, Batmobile-style scalloping scheme. Along with the beautiful new paint and accents came unique bat-shaped taillight lenses and bat-shaped spinners for the car's magnesium wheels.

The *Batcar* was a huge hit with fans and thus also became a favorite among drag strip promoters, who clamored to book Joniec for match races. Al quit his full-time job at Swenson Ford to go racing full time, and he picked up a new sponsor in the form of Rice and Holman Ford in Pennsauken, New Jersey. Ford Motor Company had plans for how they wanted Al to represent the company in drag racing for 1966 as well.

Ford told me that they wanted me run in the brand-new NASCAR drag racing series that year, instead of NHRA. Obviously Ford supported the NASCAR stock car series. NASCAR wanted to try to compete with NHRA in drag racing, so they asked Ford and Chrysler if they would help out, and both companies agreed.

As a result I would not be going to many, if any, NHRA races. At first I wasn't sure what to think about it all. However, this new NASCAR deal was supposed to be big. Because I was a part of the startup I began to feel good about it, even though I feared there would be no *National Dragster* coverage of the series, as it was an NHRA publication.

NASCAR scheduled a big series of races all though the East Coast and Midwest for us to go to that year. It worked well. Pretty much the same guys traveled from race to race, so it eventually felt like family as everyone appeared to get along just fine.

As I remember the rules were very much like NHRA A/FX, but used weight categories and

Al Joniec debuted both a new sponsor (Rice and Holman Ford) and an all new concept/look for his A/FX Mustang to start off the 1966 season. The *Batcar* was an idea that would prove a hit with track promoters and fans alike.

required that we use pump gas only. The classes were called Ultra Stock. I could barely get into the 3,200 lb. class as a NHRA-legal A/FX, so I had to lighten the car as much as I could to try and get into 2,800 lb. class for NASCAR competition. I never got that close, as the car weighed in at around 2900-plus lbs.

Cars in the Ultra Stock series competed in their individual weight class first. Then the winners would come back for a handicap eliminator race to determine the overall winner. The handicaps were based on your best time.

The handicap concept was based on the assumption that everyone had to go as fast as they could to win their weight class. The flaw was that if you could win your class without going as fast as you could, then you could sandbag. There was no incentive to go faster than it took to win, so that was something that I practiced. The car itself was basically the same as last year, just lightened.

In the fashion of the Batmobile from the Batman television series, and in deference to the scalloping on Louie Tomaselli's '58 Ford, the *Batcar* featured red scallops over black paint.

Al poses with the newly completed *Batcar* A/FX Mustang early in the 1966 season.

Among other things, I replaced all the glass with Plexiglas, removed the embarrassingly small exhaust system we had to run in NHRA AF/X, replaced the rear quarter window vents with plain Plexiglas, mounted an aluminum radiator, removed the fuel tank, and eventually put in the much-lighter fuel injection system.

The rod bearing problem from the previous year needed permanent fixing. Once Connie Kalitta was on the Ford team and running Top Fuel with a Cammer, I asked him what he was doing to avoid the problem. He said he cut the crank to the

In keeping with the theme, Al's transporter became the Bat Truck for 1966. For the time carrying a race car in this manner was considered top-flight.

If you have a *Batcar* it is only natural that you have a bat parachute to slow it down. Fans loved this finishing touch.

smaller-size Chrysler bearing and used aluminum rods. I did that as well and never experienced another bearing problem (but I did start to break rods after about 50 runs). Ford also sent me new Forged True pistons that had a very high dome that gave me 13.5:1 compression (versus the 12:1 stock pistons) and very trick Dykes top rings.

The season was a case of the good, the bad, and the ugly.

The good: I always won my class and never remember losing unless I broke something.

The bad: I was breaking more times than I should have.

The ugly: I didn't know what the hell was causing it. It was a moving target and it seemed to be many random things going wrong. I did have a good win-loss ratio in the new NASCAR series. But I did have more than my fair share of mechanical problems with the engine, and it cost me money, time, and image over the year.

I broke my first and only long timing chain. I then I started to have small timing chain problems

Ford Division News Bureau
P. O. Box 808-Dearborn, Mich. 48121
Phone:

RELEASE ANYTIME

AL JONIEC -- MUSTANG

Al Joniec, 26-year-old driver from Philadelphia, will campaign largely on the NASCAR circuit in 1966 with a 427-cubic-inch single overhead cam Mustang designed to run in the 2,800-pound class.

This will be Joniec's 10th year in drag racing, much of which has been spent in Ford products. He raced a superstock Galaxie in 1962 and 1963; a superstock Fairlane in 1964, and last year campaigned an A/FX Mustang.

As Joniec puts it:

"I'm always a bridesmaid and never a bride."

Although Al has failed to come through with a major title yet, he has compiled more than 200 trophies in his career. He holds a number of local track records as well as the Eastern Drag News elapsed time and top speed marks.

When he isn't racing, Joniec stays close to the sport in his capacity as performance manager for a speed equipment manufacturing company. He and his wife, Barbara, have two children.

This official Ford promotional document describes Al and how he would carry the Ford banner in the NASCAR drag racing series for the 1966 season.

Al and crewman Bud Rubino pose with the *Batcar.* The headlamps have been deleted and replaced with hoses that duct air directly into the carburetors atop the SOHC 427 engine.

and cam gear sprocket keyway problems on the dummy cam. This usually bent most of the valves and created all kinds of problems. I finally replaced the dummy chain and sprockets with a gear drive, which fixed it permanently. I then started to break the aluminum connecting rods, which took out the heads and valves. I found out the hard way that you had to replace the rods every 40 to 60 runs because of fatigue. The new pistons had such a high dome that when the aluminum rods stretched it started closing the spark plug gap, and that created misfires. I had to use side-electrode plugs to fix that problem. The high-dome pistons created another new problem, as the dome split the combustion chamber in half and created end-gas detonation that killed performance. To remedy this situation I trimmed one edge off of each piston dome and after that I was able to go back to using standard long-reach spark plugs again.

I also started breaking the hollow-stemmed valves, inexplicably. This was a more complicated problem for me to find. These were brand-new valves that seemed to be bad; it sometimes forced me to use old valves because I couldn't get good, new ones all the time. I sometimes had to use slightly bent ones that I ground straight. All of this was a nightmare for me.

The final blow was when I thought I must be floating the valves and dropping them that way because I revved it so high (the tach tattletale was now in 9,500-rpm range). I tried to increase the spring seat pressure. I checked the spring pressure with my new spring pressure gauge. It had a clamping stop adjuster that let me test at a fixed assembly height. When I did that I also used the spring retainers, which had a different height position of the inner spring. However, on my Cammers, I had two different sets of spring seats that came in two configurations; one had a raised center position for the inner spring just like the spring retainers, and some were just flat. Both protected the aluminum heads. I usually put the spring seats on the head because I still ran valve seals and you couldn't pass the seats over the seals.

In my attempt to follow the valve float theory, I kept reducing the spring assembly height to increase the seat spring pressure. When I checked for spring binding with the spring retainer, especially the inner spring, it was good, but not when I used the spring seats with the raised inner spring configuration (that was not a part of the assembly test). I started breaking the valves due to the inner spring binding. As a result, I lost so many valves and engines and lost so many races, it was embarrassing. On this particular spring stacking problem it was totally my fault. I always

Al's helmet on, ready-to-run promotional photo was typical for the era.

This promotional shot of the *Batcar* was handed out to a limited number of lucky fans on race day. An autographed version has been a prized memento in the author's collection since the spring of 1966.

paid attention to the small stuff as, "the devil was in the details," but I had brain fade on this one. I felt so stupid because when it ran it was a winner, but I was scared I wouldn't finish a race without breaking something. It was much later in life that I found out that knowledge and wisdom are separate things.

In the early spring of 1966, my local Philadelphia rock and roll radio stations were filling the airwaves with spots about the upcoming match race battle between Al Joniec's new *Batcar* Mustang and the altered-wheelbase Hemi Dodge of his archrival Bob Harrop, coming up at Atco Dragway in New Jersey. In my world at the time, this qualified as a must-attend event under any circumstances—not unlike a religious pilgrimage for a teenaged Ford fan like me.

On the fateful day of the awaited showdown between Ford and Mopar, two friends and I embarked on the 40-mile trip to Atco. In my usual state of financial embarrassment, but not wanting to miss history in the making, I entered the facility in a less than conventional manner—concealed in the trunk of my friend's 1955 Ford.

So special was this occasion that the first 125 fans through the gate would receive an autographed photo of Al Joniec's *Batcar.* Unfortunately for yours truly, this special offer only applied to *paid* admissions, so my friend was the recipient of the autographed photo. A short time later my friends and I were in the grandstands cheering for the Batman as he and Harrop performed their pre-run burnouts through the rosin. Al didn't disappoint his fans on this day, as the *Batcar* dispatched the fearsome Hemi Dodge in the best-of-five match race. To add to my celebratory mood, my friend handed me the signed photo of the *Batcar* saying, "This will mean more to you than it does to me." I treasure that photo to this day.

Another of Al's match-race successes during the 1966 season was reported May 1, in the publication *Drag News*, which ran the headline, "*Batcar* Flies," over its story of how Al Joniec defeated Bill Shirey's *Professor*, a Hemi-powered, altered-wheelbase Plymouth, in a match race with an elapsed time of 10.52 seconds at 133.09 mph. As of May 27, the official NASCAR drag racing series listed Al Joniec as its class record holder in Ultra/Stock at 10.62 seconds and 134.32 mph.

Yet another highlight of Al's 1966 season with the *Batcar* came when he won the Cars Magazine Championship race at Cecil County Dragway with a 9.95 second ET at 140 mph.

HOLY CHILDREN'S ICON, BATMAN, YOU ALMOST RAN OVER CAPTAIN KANGAROO!

I was asked to do some public relations work in order to promote the upcoming NASCAR series drag race in Pittsburgh for both a local newspaper and children's TV show. The premise was for me to speed up the hill at the studio, jump out with my fire suit and respirator on, and be interviewed as Batman. As I waited at the bottom of the large hill, the producer, who was standing by to give me the start signal, said that if I could make some noise and go fast it would add to the show. I said, "I can do that!"

The hill was full of dirt and dust and was very slippery, plus there had been no practice runs. When he finally told me to go I kind of went overboard. The slippery road was hard to handle with several turns going uphill. I was slipping and sliding all over the place. By the time I got to the top I was going way too fast. I slid into the area where the host of the show, a man named Paul Shannon, was standing, complete with microphone, waiting for me to step out and for my interview. Well, he didn't hang around as I came screaming up to him, sliding sideways, trying to stop. He threw his microphone in the air and ran for his life. If he had stayed where he was, I would have hit him. There was no retake, as he interviewed me standing still after that.

THE SAME MATCH RACE FROM TWO PERSPECTIVES

Al Joniec:

The most memorable race I ever had in my racing career was a best-of-three match race against Bob Harrop and his fuel-burning, altered-wheelbase Dodge Funny Car at Atco Dragway in New Jersey in the summer of 1966. This was the first time I tried to use nitro fuel (only 20 percent) and we would be using rosin for traction. We had to use fire suits and respirators as well. I felt like an

Al, Bud Rubino, and the *Batcar* took part in the filming of a children's T.V. show as a promotion in 1966. In his exuberance to put on an impressive show for the kids, Al almost ran over the show's host with the *Batcar.*

One of Al's most frequent "special guest villains" during the 1966 season would be Dodge racer Bob Harrop. Talented artist Joel F. Naprstek painted this rendition of the Harrop *Flying Carpet* altered-wheelbase Hemi Dodge.

astronaut and was really excited for this rematch of our race in the Spring.

The race had been hyped over the radio all week, and the place was packed with several thousand people in the stands. While the announcer hyped the race, all the other racing stopped. It was quiet as we pushed our cars to the starting line. We cleaned the track with a broom and put rosin down. We then both put on our fire suits, helmets and respirators. Both of us did burn outs over the rosin until I felt the tires really biting into the track.

We finally staged and I revved the engine as high as ever. When the green light came on I let the clutch out. It felt like the earth rotated while I stood still. I don't think I moved forward one inch as the car stood straight up in a giant wheelstand. I like to SHIT MY PANTS! I felt like I was in an elevator and I was looking down at the ground 15 feet below. I immediately got scared and lifted off the throttle. This brought the front of the car crashing down, bottoming the front springs like I never experienced before. Obviously I lost the first round, as Harrop was long gone. I was not sure if the car was hurt or if I could continue. The guys told me later that the crowd went wild—that was as big a wheelstand as most of them had ever seen. Me too!

We went back to the pits and checked out the car. Sure enough, the oil pan was hurt from hitting the ground, but the front suspension looked OK. We had just gotten a new experimental fiberglass oil pan from Ford, and it was now cracked and full of holes from the little rocks that went through it when it bottomed out. I got some sheet metal screws and some rubber washers to fix the holes, used 200-mph tape over the cracks, and we were good to go.

The second race we did the same thing with track prep and rosin while the announcer hyped the crowd. But this time I was ready for the wheelstand, even though I had no real plan. Just like the first race, I stood still, the earth rotated, and the car went right up into a giant wheelstand. But this time I played with the throttle to keep the car from going over backwards while still moving forward (once the center of gravity of the car is over the rear wheels in a big wheelstand, it doesn't take much power to hold it there). I made it through first gear and shifted in the air into second, still holding it in the air though second gear, when it

The *Batcar* and Bat Truck early in the 1966 season.

finally came down. By this time Harrop was ahead of me by about three to four car lengths. But as I put it in third I stopped his momentum and started to close in on him. When I put it in 4th I knew I had him, as I marched right by him before the lights. WHAT A RACE! The guys told me the crowd went wild.

The third race would have been a classic but when I let the clutch out at the start, the rear end exploded and the race was over. So, even though I technically I lost the match race, in my mind I won, and it's still the most exciting and memorable race I ever had.

Charles Morris:

The local radio stations had hyped the upcoming match race between Al's *Batcar* and the *Flying Carpet* Dodge of Bob Harrop for at least a week before the event. "Sunday, Sunday, at Atco Dragway," was the standard lead-in to get all the race fans' attention. Once again, I wasn't going to miss this race but, unlike the previously described match between Joniec and Harrop, this time I possessed the financial wherewithal to buy a ticket.

Al had his share of mechanical problems during the 1966 season. Aluminum connecting rods broke frequently, which would take out the heads and valves too. Al found out the hard way you had to replace the rods every 40 to 60 runs.

This shot shows Al's fix for the problem of his domed pistons splitting the combustion chamber and causing ignition problems. He cut off one side of the corner of the dome to form a small combustion chamber under the spark plug. This stopped the end gas detonation he was experiencing, and he could use regular extended tip spark plugs again. It worked well and the car did go faster.

Al Joniec Hopes to End Jinx

'Bat Car' Set For Drag Races

Al Joniec of Philadelphia has entered his "Bat Car" in the NASCAR Grand Stock Circuit championship races to be held at Pittsburgh International Dragway Sunday afternoon.

Joniec is among the leaders in the NASCAR circuit's point standings, but in two previous efforts at the South Fayette Township strip he has run into mechanical difficulties. He is hopeful of breaking that jinx Sunday.

He joins Melvin Yow, Lillington, N. C., and Paul Moody, Belleville, Mich., the top two in the point standings, for the full afternoon of eliminations among the top stock and ultra-stock drag machines in the nation.

Other entries already received include Carson Hyman, Camden, S. C.; Tom Sneden, Richmond, Va.; Ken Vogt and Dave Lyall, Detroit; Bud Faubel, Chambersburg; Dave Koffel, Birmingham, Mich., and many others.

AL JONIEC
A tough competitor.

This article from a racing publication of the time speaks of the mechanical woes Al suffered during the 1966 season. His racer's spirit, grit, determination and knowledge helped him turn around any bad fortunes and remain a top competitor.

Before the race the stands were packed with fans, seemingly divided equally between the Ford and Mopar camps. There was tension in the air as the announcer (I believe it was Bob Frey) continued to harangue the crowd and the two teams prepped the track with rosin. Personally, I loved this part of a match race. Watching the combatants burn through the "gold dust" and then back up into the tracks left by the rear tires really got my blood up.

The *Batcar* had undergone obvious modifications since the spring match race, but it was Harrop's Dodge that really looked radical. While the *Batcar* still looked like a Mustang you could find at your local Ford dealer, the Mopar was the epitome of what gave rise to the term "Funny Car." With its radically altered wheelbase, acid-dipped body, and long-stack fuel injectors protruding from the top of the Hemi engine, Harrop's car looked other-worldly to me.

When the lights came down for the first race the *Batcar* went straight up in the air. The Dodge got out of Dodge, as it were, taking a huge lead. I paid little attention to the Mopar at this point as the *Batcar*came crashing down to the track. I can distinctly recall commenting to a friend, "Did you see what he just did?" Remember that match racing at the time was very much about pageantry and the show, so my perception was that the Batman had planned this entire episode!

The announcer indicated that the *Batcar* had suffered some damage, but Al and Bud Rubino, whom the track announcer referred to as "Robin," were hard at work trying to make repairs for the next round. I don't mind telling you that when the *Batcar* went sky-high off the starting line in the second matchup, giving Harrop a huge advantage yet again,

This photo of Al unloading the *Batcar* from the transporter at Atco Dragway early in the 1966 season is believed to have been taken on the day of the first match race with Bob Harrop, which was the day I received my treasured autographed photo.

I was worried. Imagine my delight when the Batman overcame adversity to run down the fleeing Dodge and light the win light in the Ford lane. We went crazy!

As Al came up the return road in front of the grandstands, thousands of fans lined the fence, cheering for him. As Al indicated, his car broke in the third stanza, giving Harrop the match. But my friends

Competition evolved rapidly in 1966; in order to keep pace, Al made some modifications to the *Batcar*. Most evident was the addition of fuel injection.

The *Batcar*'s SOHC 427 engine with the fuel injectors in place was impressive enough to be featured on the cover of *Hi-Performance CARS* magazine in 1966.

A reference to the campy Batman TV series was cleverly carried over onto the stacks of the fuel injectors, which were adorned with the graphics that punctuated fights between the Caped Crusader and his featured villains.

and I left the track duly impressed by our beloved *Batcar* and its killer Cammer engine. Years later I got to know Bob Harrop and found that he is a great guy. I confessed that I was one of those Ford fans who booed him from the stands so many times. I apologized, saying that had I known what a nice man he was, I would have felt guilty. These were the times when brand recognition was so very important to race fans like me. By wheeling that factory-supported Dodge, Bob was considered the enemy.

The *Batcar* also went on a diet during the 1966 season. Al got the weight down to a point where he could run in the 2,700-pound Gas class in the NASCAR drag racing series.

The 1966 Super Stock Nationals

At this race I won my class (2,700-pound Gas) so I was eligible to run for the top eliminator. I had broken my Ford Toploader transmission in the process of winning the class. I had never broken a Toploader before, so I never thought to get an extra one. Instead, I put in my old T-10 that I kept as a spare. Because the first gear was higher in the T-10 and the sun and the moon were aligned, the car not only went even faster but did huge wheelies as well.

All the Top Eliminator competitors ran against each other via a handicapped start, and the handicap index was set off of your fastest time. Because I had easily won my class, and the car was now even faster with the T-10, I was kicking ass and taking names. It was down to the final four cars for the overall eliminator and I was racing Ed Schartman (with his unlimited flip-top

Of Al's many match race victories with the *Batcar* in 1966, perhaps the sweetest of all came when he met and defeated the *Black Arrow* Plymouth of "Doc" Burgess and Bill Jenkins. He looked upon this as payback for Jenkins "firing" him at the beginning of 1965. Atco's advertisement for this match race announced that Joniec was undefeated in match race competition to that point. He remained that way!

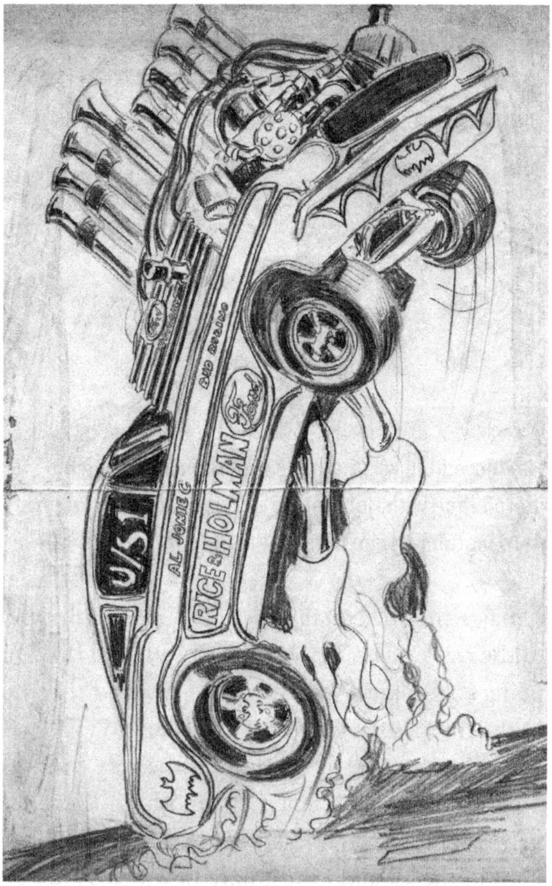

Following the day that Al stood the *Batcar* on its back bumper during a match race against Bob Harrop at Atco Dragway, a friend drew this cartoon to commemorate the event.

fuel Funny Car) when my life would change yet again. Because of the handicap I was well ahead of him, but during the race my clutch pedal was moving over to the side as I shifted gears because the retaining clip holding the pedals to the shaft fell off. By the time I shifted to 3rd the pedal had moved over so much that it missed the return position stop. I had to lift my leg way up for the next shift and was losing valuable time shifting trying to figure it all out. I did finally get it into 4th gear, but he just flew by me right at the finish line. I thought I most likely would have won if I hadn't wasted so much time with the clutch.

Little did I realize that the shaft was shared by both the clutch and brake pedals, so both were affected by the retaining clip's disappearance. I went through the lights at 144 mph and when I went to push on the brake pedal it went right to the floor. The clutch shaft had moved over far enough so that it disengaged the brake pedal. I started to panic so I pulled the seldom-used parachute release. When I didn't feel the car slowing I looked back and there was the parachute doing the Roman Candle thing—it was out, but not open.

So there I was with no brakes, no clutch, and no working parachute. To top it all off, in my panic I had tried to shift into a lower gear, but because the clutch linkage had now fallen off, I had no clutch and had only succeeded in getting the transmission into neutral, with no hope of getting into any gear to help slow the car down. The end of the track was coming up quickly.

I clearly remember seeing Jack Chrisman's fuel Funny Car sitting on the side, still smoking from

The incredible wheelstands experienced during his match race with Bob Harrop's *Flying Carpet* prompted Al to install wheelie bars on the *Batcar.*

The *Batcar* was always a fan favorite wherever Al raced. Here a young fan does his best to get a peek at the awesome Cammer engine that powered the car.

a major fire that totally destroyed it near the end of the track, as I said to myself, "HOLY SHIT, I'm going to get hurt!" By the time I reached the end of the track I estimated that had I slowed to somewhere between 70 to 90 mph. At the very end was a 10-foot chain link fence. When I hit it the fence bent over and acted like a ramp. It launched me high in the air and I must have flown close to 100 feet. I was so tense I straight-armed the steering wheel and straightened my legs as hard as I could.

It felt like forever as I shut my eyes and held on for dear life, waiting for the car to finally stop flying. It hit the ground so hard in the front that I couldn't believe it. It acted like a springboard for the entire car. The car rebounded at least 50 to 60 feet in the air and tumbled end-over-end, wrapping the parachute around the car. It flew over 40- to 50-foot-high trees and landed on the other side of them.

When the car finally came down (I was still hanging on for dear life and had my eyes closed tight), it landed on all four wheels at once with a huge crash. It bounced up but kept going (speed

CLASSES
SUPER STOCK
INTERNATIONALS

27G CLASS DESIGNATIONS

The first of the gas classes is the 27G class. This is very similar to the 27F class with two important differences. One, the cars must run gasoline for fuel. Two, the displacement of the engine is limited to 430 cu. in. The weight range, from 2700-lbs. to 2999-lbs., is the same. The same 15% engine relocation is allowed. No blowers are permitted.

This class will find in attendance many of the 2700-lb. cars that have been competing on the NASCAR circuit. Several of last year's MoPar funny cars are expected to be in this class and should add plenty to the overall color and excitement.

Keep an eye out for times in the high nines to the mid tens. Speeds should creep over 140 mph.

The NASCAR 2,700-pound Gas class rules featured the *Batcar* as an example of the class standard.

As the 1966 season was winding down, Al entered the *Batcar* at the Super Stock Nationals held at New York National Speedway on Long Island. In this photo he defeats Dave Koffell's Plymouth.

doesn't kill, it's the sudden stop that does it) for another 75 to 100 feet, hit and uprooted a smaller, 10-inch-diameter tree, and finally stopped. I couldn't believe, after all of that, I wasn't hurt bad and walked away. Actually, I ran away—I couldn't get out of the car fast enough (the door still worked), as fire was now my biggest concern. From outside the car I looked at it all like a third-person observer and it was surreal.

I had big bruises on my shoulders from the harness and a very big one on the center of my chest from the steering wheel shaft. From straight-arming the dished steering wheel, I had bent it completely back the other way. I think it saved me by doing that because the steering shaft would have killed me if it had gone another inch or so. I also straightened my legs while pushing on the clutch pedal, which I felt pushed me back in the seat hard enough to save me from far worse physical damage when the car took the huge hits that crushed the roll cage tubing. I only had some bruises and my body was sore all over, but I walked away.

The front end of the car was completely totaled from the several big-time hits it took, but the back was still OK, as even the doors still worked. I didn't hang around after that as they took me in the ambulance for observation. I was released soon, after the tests they did and my complaining convinced them I was OK. I was lucky that my best friend Jim Rahm had flown his airplane to the race track. I flew home with him right away. The guys told me later that they had to cut down trees and dig out the car, which was well into the woods, and somehow load it (the front wheels were useless) onto the truck late that night. Years later a photo in *Hot Rod* magazine showed the wrecked car with the parachute wrapped around it, supporting my recollection of flying and tumbling.

1967: The Batcar *Stretches its Wings (and Nose)*

After we got the car back from the wreck at the S/S race, we looked it over and found that everything from the firewall forward was junk, but the basic body was useable. I thought it was time to make it like the other Ford team cars—

Another shot from the same day at New York National shows Al in the *Batcar* taking on Doug Nash's *Bronco Buster.* The *Batcar* was flying at this event; little did Al know that in a short time he and the car would fly—literally.

This grainy clipping shows the crash site well of the end of New York National Speedway where the *Batcar* finally came to rest after knocking over a fence, flying through the air, clipping some trees, and miraculously landing upright.

SOME SLOGAN – Al Joniec's trunk slogan "What A Way To Go-Go" proved too true. He charged off line, then continued off strip and into distant woods, extensively damaging Mustang.

This clipping shows a closer view of the wrecked *Batcar* with its parachute completely wrapped around the body. This occurred during flight and proved Al's account to many doubters.

1967 dawned bright with the *Batcar* getting an all-new, extended-wheelbase, engine set-back look. This snapshot was taken at MAI just after the completion of the rebuild for the 67 season.

The official promotional photo of the *Batcar* shows off the extended front clip and fresh black paint with diamond dust and candy apple red scallops.

Atco Dragway once again featured an autographed souvenir photo of Al and the latest incarnation of the ever-popular *Batcar*.

a lightweight, altered-wheelbase car—by moving the rear forward and extending the front end. The fiberglass front end was already available from Ford.

The wrecked car was at Jim Rahm's place, MAI (Manufacturers Associates Incorporated) a speed equipment warehouse in Hatboro, Pennsylvania.

The *Batcar*'s Hilborn fuel-injected, SOHC 427 engine was back for 1967. Because it was now backed with a C-6 automatic transmission it would prove more reliable and deliver more consistent elapsed times.

MAI was located in a shop across the street from big-time modified dirt-car builder named Paul Deasey (his car was called *The Big Donkey*). When I first went to MAI, Paul came over and introduced himself. He was a Ford guy and was really interested in the Cammer. Because he built cars from scratch and was across the street, he won the job of making my new rectangular-tube chassis for the next generation *Batcar*. The deal was based on the barter system. He would do the car to my specs and in return I would give him a Cammer engine for his dirt car. Paul was a good man and we became friends; I got my car done the way I wanted it, and he went on to win a lot of races with the Cammer. It was a good situation for both of us.

The big change in the powertrain department was a C-6 automatic transmission. This would make for huge improvement in engine reliability, as it eliminated overspeed situations such as broken driveline parts and/or missing shifts. I tested the car at Cecil County and it went straight and smooth, turning low nines at 150 mph on gas the first time out (conditions were perfect to go fast).

5,000 Pack Maple Grove to See "Batcar" Ford Strike Down Foe

Al Joniec (near lane). took the measure of two opponents at Maple Grove with his "Batcar" Mustang. Al turned in the low nines and 143 mph to top the "Kount Hemi" Plymouth funny car before over 5,000 fans at the Maple Grove opener.. Originally, the Mustang was to face Bud Faubel and his Dodge, but Bud broke on a trial run.

–photography: Jim Cutler

READING, PENNSYLVANIA (Maple Grove Park Dragway) April 2 – Sunny skies and eighty degree temperatures greeted spectators and racers alike as Maple Grove Park Dragway kicked off the 1967 season. A crowd of more than 5,000 people turned out to witness a return match of Al Joniec's Phila. based "Bat-Car" Mustang and the Faubel and Weiler Dodge Dart, "Honker," from Chambersburg, Penna.

The match came to a standstill when Bud Faubel broke a camshaft in a warm-up round at 10.23 elapsed time and 137.64 mph. Promotor Bob Eveland substituted Allentown's E. J. Lilly with the fuel burning "Kount Hemi" Plymouth funny car, and the match race was on.

Joniec's "Bat-Car" proved too much for the fuel burner, running elapsed times in the low 9's at speeds better than 142 mph. Joniec will spend much of the season on the NASCAR Grand Stock circuit and will be a strong favorite when the circuit swings to Maple Grove Saturday night, June 10th.

A surprise visitor today was Frank Monaghan, new owner of the "Gemini Cricket." Frank, who hails from Phila., Pa., delighted the crowd with some twenty foot wheel stands in his first attempt at the controls of the Dodge truck. Also on display in the pits, was the new and beautiful 427 powered Camaro, "Think Twice," of Philadelphia's Jim Keese.

Maple Grove successfully inaugurated the Class or Eliminator system to the delight of the packed stands, as the fans were able to witness the hotly contested eliminator action early in the program.

This new system adds much spectator appeal to the sport of drag racing. Competition Eliminator went to Jack Hilfirty's C/D over Don Westis. Jack's hole shot was more than enough to hold off the top end charge of Westis' quicker e.t. of 10.06 at 143.31, good enough for top time and low e.t. of the event, to Jack's 10.23 – 141.73.

Reading's Ed Showalter and his B Stocker Chevy coupled a 12.87 e.t. with a 109.09 mph to edge out a top stock win over Glen Finacle's Pink Panther AA/SA Dodge from Willow Grove, Pa. Junior Stock saw Charlie Lutz's D/S "Little Terror" Chevy overcome a tremendous handicap to best Lou Cooper in his P/S Chevy.

Lutz went 13.24 at 102.71 mph to Cooper's 18.78 – 73.05. Both cars hail from Reading, Pa., and make their presence felt to all the Junior Stockers wherever they appear. Perhaps the hardest fought competition of the event was in the Street Eliminator bracket and two of Reading, Penna.'s finest competitors battled their way through to the final round.

Bechtel overcame a slight handicap and won a close one in the traps with a 12.11 e.t. at 109.73 mph to Al's 12.87 – 105.05.

The new long-nose *Batcar* came out swinging early in the season, as can be seen here in this news clipping and photo. 5,000 fans showed up at Maple Grove Raceway to see the caped crusader put the hurt on the *Kount Hemi* Plymouth.

The big difference was that the 5,000 rpm stall-speed torque converter made for great launches off the line. It felt strange using an automatic (kind of like cheating, it was so easy), but I didn't mind after I saw the time card.

I was as happy as I could be with my clothes on and was looking forward to the first race of the year in Deland, Florida. It was kind of like the East Coast version of the Winternationals in California. Some of the eastern Ford teams, and Ford management, where there. It was a big NASCAR deal, with more cars in my class than ever before, as the Chrysler teams were there too.

Normally Sox & Martin didn't go to NASCAR races, but they were at this one and easily had the fastest Mopar that day. I had the fastest car overall, though, and after running through a full

The popularity of the *Batcar*, along with Al's winning reputation, brought sponsors calling. Crane Cams used the *Batcar* as a feature in its 1967 ad campaign.

More often than not during the 1967 season the Chrysler Hemi-powered competition fell victim to Joniec's *Batcar* thank to its potent SOHC 427 engine and the consistency of Al's driving. Here Pee Wee Wallace drops another round to the mighty *Batcar*. Losses like this one would send the Chrysler camp back to the drawing board for 1968. The secret to MOPAR's successes against Al in 1968 would not be revealed until it was too late. (Lloyd Wolfe)

A smiling Al Joniec sits in his "office" and confers with some well wishers before taking on his next "special guest villain" in NASCAR Ultra/Stock competition. (Lloyd Wolfe)

Atco Dragway Victory To Joniec's 'Batcar'

Al Joniec piloted his "Batcar" Mustang to a sweep in the Atco Dragway feature yesterday, topping Pee Wee Wallace's Ultra-Stocker in three straight races.

Joniec drove his '67 Mustang, powered by an experimental 427 cu. in. overhead cam, Hilborn-injected Ford engine which develops up to 700 horsepower on gasoline, to victory with elapsed times of 9.75 (143 m.p.h.), 10.17 (139) and 9.67 (142).

Wallace's best time in his '67 Barracuda, which has an all fiberglas body, uses an especially designed tube chassis and receives it's horsepower from a 480 cu. in. Golden Commando Hemi engine, came in the third race with a 10:19 E.T. (130 m.p.h.). His other times were 10.80 (127) and 10:83 (117).

Top Stock

Joniec's Mustang is from the Rice & Holman Ford Agency in Merchantville.

Maurus Petruzzi of Philadelphia captured top stock honors with a 17.00 E.T. (78 m.p.h.) in his '51 Oldsmobile to triumph over Jim Lane of Mount Holly.

Audubon's Eric Ferddler drove an AA gas dragster to an 8.90 E.T. (179 m.p.h.) for top elminator. The super-charged Plymouth-powered dragster is owned by Joe Girdleman of Hainesport.

Batman Tops in Best of Five

PITTSBURG, PA --Zap! Pow! Biff! Bam! Bring on the Riddler, the Joker, the Mad Hatter and all the rest. (Ed Note: Bring on their girl friends, too) We're ready for all of them. My talented, super smart boss, The Batman, just cooled off Vernon Rowley-Al (Buckeye) Phillips and Co. here in a best-of-five series.

Of course we figured we had all the answers after dealing with that crew we tangle with all the time on television. But this Buckeye-Vernon team was something else.

First they pulled a Joker trick on us-by making it rain. Thus, it was a mighty damp Batman, alias Al Joniec, who had to go against them.

Then the announcer at Pittsburgh pulled a Riddler trick on us-by saying we'd race against Vernon Rowley. So Al, oops, make that the Batman, proceeded to take a big dose of his 'Beat Rowley' special hate capsules-only to have Al Phillips climb in behind the wheel to do battle with us.

The final clincher was when we had trouble finding the Batman's famous bat helmet-obviously a foul trick of the Mad Hatter.

For a while, it looked as if they'd have to tune in here next week to get the final result but the rain finally halted (I doubt if there was anything left up there to pour down anyway) and get this: Even with all of the obstacles, my super ultra boss, in his super ultra Mustang Bat Car, won three of the four runs.

Batman won the first two-then, of course, we had to slow up on the third so that the cameras could get Batman's picture smiling as he went by so Buckeye's Barracuda won that one. But then Batman turned it on again strong to win the final run.

More news clippings show just what a good year Al and the *Batcar* were having.

BELOW LEFT—Holy Hilborns, Batman, it's an impostor. The East Coast version of Batman, Al Joniec, fights a perennial battle against time, instead of crime. His weapon? An SOHC Mustang, sponsored by Rice & Holman Ford. BELOW—The revamped Batcar has a metalflake black paint job, finished off in cherry flake scallops. BOTTOM OF PAGE—Gene Collins (right) of Washington, D.C., whipped it on all comers in Stock Eliminator. His D/SA Mustang beat the Williams and Moore GTO 13.05-114.64 to 13.38-97.82.

Another of the *Batcar*'s match race victims during the 1967 season was factory Dodge driver Bud Faubel in his Hemi Dart Funny Car. The match took place under the lights at Maple Grove Raceway. (Lloyd Wolfe)

Al Joniec's *Batcar* versus his main protagonist during the 1967 season, Pee Wee Wallace with his *Virginian* Hemi-powered Barracuda Funny Car. This action took place at the Super/Stock Nationals in Cecil County, Maryland. (Lloyd Wolfe)

Al is seen here at speed in the *Batcar* during a run down the track at Cecil County Dragway during the 1967 Super/Stock Nationals. (Lloyd Wolfe)

field of cars, it was Ronnie Sox and me for the trophy run. Up until that time I was fast idling on the starting line and just floored in when the light came on. This approach let the converter snap up near to its 5,000 limit, which was like letting out the clutch on a manual trans and was easier on the equipment. Even though I had Sox covered by at least two tenths of a second, I really wanted to beat him badly in front of the Ford brass, so I stood on the brakes to take the converter to its max 5,000 rpm in gear, in an effort to really blast off the line. This was faster ET-wise, but you could watch the transmission get red hot, as the temperature climb from the high-stall converter was so fast it was hard to believe.

As I approached the max rpm, the car moved sideways through the lights before the green light came on and it was all over for me. What happened was something I never saw before or since. The left end of the rear housing simply twisted around like a pretzel from the torque (the brake held it at the end and the engine torque twisted it). After I got the car back home, I noticed the trans oil was burned as well. I pulled the transmission and sent it back to Ford for an exchange. I put the replacement in and was ready for the next scheduled NASCAR race soon after that.

When I went to the track to test the car with the new rear and transmission, I couldn't believe how much the car slowed down—three to four tenths, and it was all off the line. I couldn't do a wheelie no matter how hard I tried. Apparently, the first trans I got had a good working stator in the converter, but the replacement didn't. The stator multiplies the torque at low speed and acts almost like another gear. All the team Funny Cars had so much power they didn't need nor want their stators to work because they would blow the tires off. So, as far as I know, Ford took out the stators in all the transmissions in the rotating pool.

If you remember I said there was no incentive to go any faster than required to win because of the handicap system we had to run in. So I started winning my class using this transmission without the stator, so I kept my mouth shut, knowing that if I had to go faster, I knew how. Bottom line, everything was now so smooth I just didn't break parts like I did last year. As soon as I got into the rhythm of the regular season I was not only

The *Batcar* on the starting line at Cecil County Dragway. Known as the "traction capitol of the East," Al had great success at this track over the years. (Lloyd Wolfe)

winning my class but I was winning a lot of the handicap eliminators as well.

Bob Ramsay, writing for *Drag World*, had this to say about the return of the *Batcar* to competition:

> "One thing was certain, the Batman has lost none of his following despite his short absence from the match race scene. Al Joniec has really done a lot with what began as a gimmick idea related to *Drag World* last winter. His successes should serve as an example to those who are willing to combine thought and hard work to come up with a winning, crowd-pleasing car, one that is a standout in what is fast becoming a nearly overpopulated class."

All American Drags Illustrated magazine reported on the NASCAR Great Lakes Summer Championships event as follows:

> "All the way from Pennsauken, New Jersey came Al Joniec and the Rice and Holman Ford Mustang. The Caped Crusader did it to everyone at one time or another during the meet and ended up winning both the Heads Up and Handicap Eliminator. Joniec won with runs of 9.85, 9.82, 9.82, 9.73, 9.90 and 9.74; seven runs within .17 seconds would indicate consistency right? While everyone else was red lighting or running out of their respective brackets, Joniec played it cool and relied on top-end charge to win."

At this event Al defeated Mopar drivers Pee Wee Wallace, in the final for Heads up Eliminator, and Tom Sneden, for the overall win in Handicap Eliminator. A photo of a race between Joniec and Wayne Gapp from the event has a caption that reads, "Wayne Gapp's *Super Cat* (Cougar) leaps off the line only to have his claws clipped at the big end by the Caped Crusader."

Here's Al looking over the tops of the *Batcar*'s fuel injector stacks while waiting for the next round of competition to begin. (Lloyd Wolfe)

Al Joniec, at the 1967 Super/Stock Nationals.

A closeup photo of those famous injectors atop the *Batcar*'s SOHC 427 engine. The campy, Batman-styled lettering made these some of the most-photographed injectors in drag racing. (Lloyd Wolfe)

The *Batcar* on the starting line at Maple Grove Raceway on a fine day in 1967. The negative of this photo, and many others shared here by Lloyd Wolfe, would go un-developed for five decades.

Yet another magazine scribe wrote of the NASCAR Grand Stock event held at Suffolk, Virginia, "Joniec's powerful S/US 1 Mustang was not to be denied in the big No Handicap Bash!" Winning time for the *Batcar* on this day was 9.71 at 145.54 mph.

At one race somewhere around the Chicago area, the promoter had both the full NASCAR schedule race (cars running on gasoline only) that we normally ran, and he added a local Fuel car circuit that used nitromethane. The traveling NASCAR official, Frank LeSueur, who ran the show, was both surprised by this and at the same time looking forward to see who was faster. I won the NASCAR race and the promoter asked me if I would race the winning car in the Fuel circuit. I said, "Sure." It was a real cliffhanger as I gave him a hole shot and held him off in a real close race to the finish line. Frank LeSueur was happier than me about it all, though I must say it felt good.

One of Al's most notable victories during the 1967 season came on June 24, when he took the *Batcar* to victory in the 2,400 pound Gas class at the Super Stock Nationals in Cecil County, Maryland. The Batman stopped the clocks at 9.63 seconds and 142.65 mph for the class win at one of the East Coast's premier events of the year. He followed this up by winning his class again at the *Drag Times* East Coast Nationals at Atco, New Jersey on September 17, defeating Dave Lyall's Fairlane in the final round with a 9.57 elapsed time at 144 mph.

Near the end of the season I was racing my main competition, Pee Wee Wallace, who I had beaten regularly all year, but when I raced him this time he was at least two cars ahead at the far end. I said, "Holy shit, here we go. No more easy going," and I had to start working hard again. I put in a new clutch C-6 , which was an automatic transmission converted to use a conventional clutch for launching the car. The next time we raced I was much faster. Our race was a cliffhanger, but I still lost by a hair. These two lost races also caused me to lose the season championship by only a few points.

Years later I met Frank LeSueur at a car show and we started talking about the good old NASCAR circuit days when he told me how Pee Wee started beating me all of a sudden. Apparently Chrysler was so pissed off that I was winning so much, the factory provided an engine for Pee Wee. They built him an oversized, 486 cubic inch engine, and the way they fixed it so it would pass tech inspection was brilliant. Every race all of us had to remove our spark plugs during the tech inspection for a P&G displacement test. The P&G test equipment was used by tracks to measure displacement. Normally just one cylinder was checked. Apparently the Chrysler boys drilled 2 very small, strategically sized bleed holes right along the spark plug threads. When the tech guys screwed in the P&G adaptor, the holes were exposed and bled off just enough of the cylinder volume to make it appear it was smaller than is really was. In other words, their 486 cubic inch engine would appear to be a 426. Again these were strategically placed, as Chrysler engines used spark plug tubes instead of washers. The tubes were large enough to seal the bleed holes when tightened. Clever as hell, but the info came too late for me to do anything about it.

As I recall these stories, I realize that not only did my car look good, but it was popular at the track and good for the NASCAR circuit public relations as well. I took all of the hard work and drudgery of the previous year and kept working hard (and learning) to make it all happen, converting what I had learned into making it fun. Bottom line, the car was not only fast, it was reliable, something I hadn't experienced the previous year —it was my winningest car. I can now clearly say that out of all my cars, the time spent racing the *Batcar* was easily my favorite.

In its September 1968 issue, *Hot Rod* magazine had high praise for Joniec's 1967 season, reporting that he had won 11 of the 16 NASCAR drag racing series events, making him the "top money man."

CHAPTER SIX

1968: THE *BATCAR* PUTS ON A FUR COAT AND AL MAKES HISTORY AT POMONA

UPON SEEING THE *BATCAR* covered in a coating of electrostatically applied blue flocking and sporting a new name, *Hairy One*, at the beginning of the 1968 season, my reaction as a spectator went something like, "Holy *&%! Batman put &*%# fur on the *Batcar*!"

In my opinion the *Batcar* had to have been the coolest concept for a racecar ever. Seeing this rendition of the beloved Mustang broke my heart, to say the least. The truth of the matter is that racers are always looking forward to the next thing, as they should be. When Al's friend Jim Rahm became a distributor for a new, revolutionary automotive finish called Velvetex, it made perfect sense to showcase the product on Al's racecar. Al has also informed me on more than one occasion that the flocking did nothing to slow the car's terminal speeds. All that was impacted was the car's appearance and my feelings.

As the *Hairy One*, and with an Art Carr clutch C-6 transmission (an adaptation of a manual transmission-style clutch to an automatic transmission, which allowed the driver to have more input and control over the car's launch off the starting line), Al's Mustang returned to do battle for one final season in the NASCAR Ultra Stock drag racing series. By this time other competitors in the series had begun to field full-chassis, flip-top Funny Cars, along the lines of those run in the NHRA, which were much lighter. It wasn't long before Al found himself at a disadvantage in a series where he had once dominated. As was always the case in the golden age of drag racing, the *Batcar/Hairy One* would soon be relegated to outdated status.

Things were changing in NASCAR racing. I did get beat by Pee Wee Wallace's Mopar at the very end of the 1967 season, so I was worried that things would not be the same anymore in 1968. There was also a flip-top Funny Car running in my class now, although still running on gas.

The Chrysler teams had not been sleeping over the winter and had found a lot of power for their Hemi engines. I could no longer dominate the class with the combination I had. Ford wanted to get out of the Funny Car business and really never did any further development work on the SOHC, leaving it all up to the individual racers. Even the few Top Fuel and Funny Cars that were using the SOHC were starting to abandon them. I felt like I was running against the Mopar company all by myself. I was running a manual-clutch C6 transmission off and on that worked great, but it only had 3 gears. The torque converter acted like a 4th gear when the stator was working, but there were reliability problems. My car was working well, but as I was standing still. All of the Chrysler cars were not only catching up, they

For the 1968 season Al changed the look of the *Batcar* by electrostatically applying a product from Jim Rahm's MAI called Velvetex. The blue material had a fuzzy texture and was not shiny like other automotive finishes. This photo was taken just after the car's completion.

were passing me. I did win my fair share of races, but the handicap cushion was now gone, and I just didn't win like before. I did OK in the beginning of the season, but as the Chryslers figured out their combinations I began to lose regularly near the end of the year."

With that sad news out of the way, we can move on to another, far more positive happening in the racing career of Al Joniec. It was one that would catapult him, the Ford Motor Company, and a new engine offering from Dearborn into the record books and onto the front pages of automotive publications overnight.

During the winter of 1967, I got a call from Chuck Foulger, who was Ford's drag racing coordinator at the time, about racing the new Mustang called the Cobra Jet. Ford was going to introduce it at the 1968 NHRA Winternationals in Pomona in February. He said that Ford wanted to go back to the, "race on Sunday, sell on Monday" concept, designed to sell cars off the showroom floor and not crazy Funny Cars. He invited me to be a part of Cobra Jet team if I wanted. Because I was not interested in Funny Cars per se, I thought this was a great idea, and I accepted the offer.

Along with the new finish came a new name. The beloved *Batcar* was now known as the *Hairy One*. While the new look was unique the *Hairy One* did not attract the fan base that the *Batcar* had.

As the time got closer we talked back and forth regarding the logistics of how we were going to make it all happen. We decided that I would take my old *Batcar* (now the *Hairy One*) and an empty trailer (to bring the new car back with) and meet at Holman-Moody West (a.k.a., Bill Stroppe

The transporter that carried the *Hairy One* to and from races also received a new look for 1968. While painted blue, it did not receive the Velvetex finish.

Engineering in Long Beach, California) with the other Ford team members. I told Chuck I would be bring my artist, Ralph Hart, to letter the car. I did this so Chuck would leave the paint on my car alone; this was the reason my car looked different from the other team cars, which were all lettered by the same guy.

Because Holman & Moody West was across the country, this trip looked to eat up about 10 days. To reduce the time away from home as much as possible, we left as late as possible, which forced us to drive for 56 hours nonstop from Philly to California. We rotated three drivers, Bill Boyce, Ralph Hart, and me. We used the back of the crew cab as a sleep area while the other two rode in the front. We only stopped for food, fuel, and to use the restroom.

We arrived in California and started right in, as all of the other team cars had been prepared and had already been lettered. It felt like we had arrived late to the party. I brought a new camshaft

The *Hairy One* Funny Car traveled to Pomona from Pennsylvania, where it was entered in competition at the NHRA Winternationals, along with Al's new Ford team 1968 1/2 Cobra Jet Mustang. Racing both cars proved to be too much, so the *Hairy One* sat out the class eliminations.

that I worked out with Crane, and that turned out to be a good idea. Ford had rented a race track for testing as well, and H&M would transport all the cars to the track, so if your car wasn't ready, you missed the test session.

Quite frankly the hangover of nonstop driving, plus working our asses off prepping the car, was tough. It also seemed that most of the other guys who were there acted as nothing more than ceremonial drivers—most of them viewed the transition from an eight-second funny car to an eleven-second Super Stocker as a step backward, with the exception of Hubert Platt, who was working his ass off just as I was.

The test sessions were private, meaning that the times were not announced, and were available

Both the *Hairy One* and Al's new Cobra Jet Mustang would travel 3,000 miles back to Pennsylvania in this configuration, with the Funny Car on the ramp truck and the Super/Stock car on the open trailer behind it. The team would narrowly avoid disaster on the long tow home.

The *Hairy One* spent most of the time in Pomona resting in the pits. The Velvetex finish did attract a lot of attention from the west coast fans according to Al.

Back on the East Coast the *Hairy One* picked up where the *Batcar* left off in NASCAR Ultra/Stock competition. Here Al is seen launching the car at Suffolk Raceway in Virginia.

Al Joniec backs up on his rubber marks just prior to a run in the semi-final round of the A.H.R.A. Grand Stock Meet at New England Dragway. Joniec had top speed of the gas funnies at 146.34 mph.
– photo by Paul Wasilewski

While the competition in Ultra/Stock had gained ground, the *Hairy One* still had some moments in the sun during the 1968 season. This racing newspaper clipping announces that Joniec's 146.34 mph blast was good for the top speed honors at the AHRA Grand Stock event held at New England Dragway in Epping, New Hampshire.

only to each individual car. I mostly worked on the suspension, as the tires were still not that great, and traction was not the best either because of the car's heavy engine and short wheelbase. Loosening the front shock rubber mounts seemed to help a lot. To make the minimum weight limit, I drilled the back brace part of the roll bar and poured buckshot into it, which added around 25 lbs. in the rear. We did get some full baseline test runs in, which were around the mid 11s. The next day I decided put in the Crane camshaft. This really made a difference, as I picked up at least two to three tenths and was now capable of running low 11s. Chuck Foulger told me at the end of the day that I had the rest of the team cars covered by at least two full tenths. We could easily run in the low 11s, which was more than seven tenths of second below the current NHRA class record, so we were in good shape. With that I called it a day and left Ralph Hart to do his lettering on the car for the race.

"We slept in the following day as we all needed a much-deserved rest. The next day was the first day of the Winternationals. I thought the car really looked really great and we went through NHRA tech easily. It felt good to be ready in advance and at ease for a change. Chuck Foulger, however, was nervous as hell, as he was responsible for this huge effort Ford was putting together. He was the drag race coordinator at Ford and this was the first big deal under his leadership. He not only had to deal with Ford, but he also had to sell and deal with the politics of the NHRA, as this was a new car that was planned for production. However, none of our cars were production cars—they were all hand-built. He had to give the NHRA a true, stock-configuration version of the vehicle to drive to show credibility (it too was hand-built out of the first 10 of the 50 bodies-in-white).

While Hubert Platt had actually unveiled the first factory-team Cobra Jet Mustang, running in the C/Stock class at the AHRA Winter Nationals in late January, 1968, the official debut of the cars was at Pomona. At Pomona the factory team cars were driven by Gas Ronda, Don Nicholson, Jerry Harvey, Hubert Platt, and Al Joniec. Ronda and Nicholson's cars were equipped with automatic transmissions and ran in the SS/EA class. Ronda would lose to Dick

At the end of the season Al sold the *Hairy One* to Drake Viscome of Carmel Ford in New York state in return for cash, a used car for his dad and his new best buddy, Butch the Bulldog. Drake continued to campaign the car on the East Coast under the Carmel Ford banner.

Here is a copy of the original order with the VIN list for the Ford factory team Cobra Jet Mustang's to be delivered to H-M Stroppe for preparation in advance of the NHRA Winternationals.

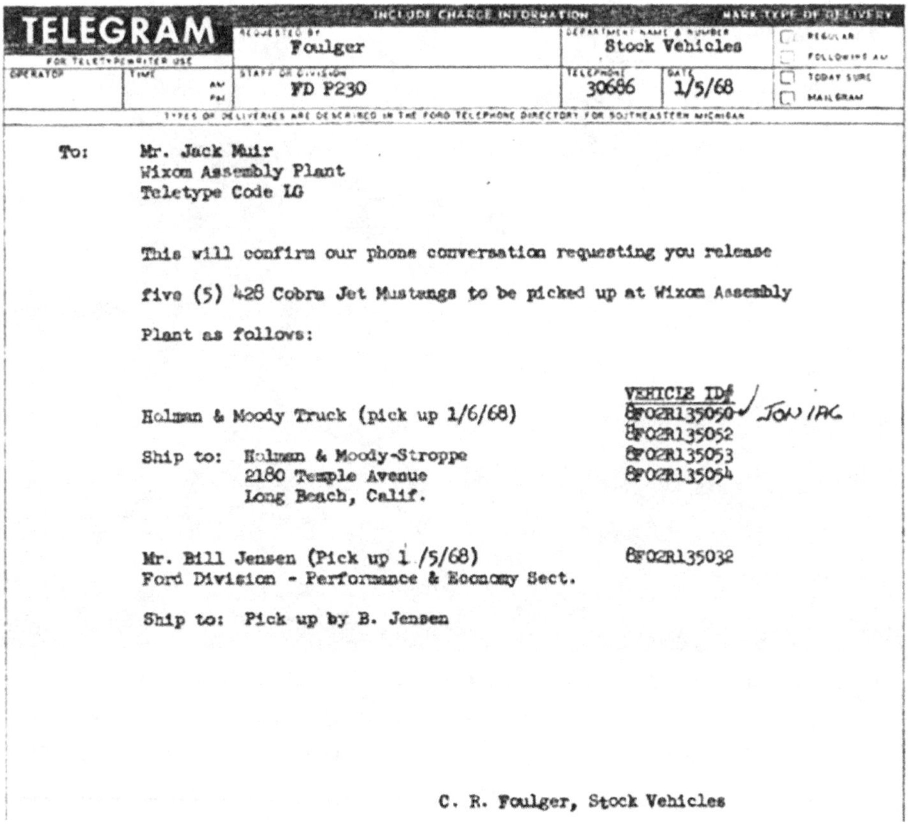

TELEGRAM

FOR TELETYPEWRITER USE

INCLUDE CHARGE INFORMATION — MARK TYPE OF DELIVERY

OPERATOR	TIME	REQUESTED BY	DEPARTMENT NAME & NUMBER	
	AM / PM	Foulger	Stock Vehicles	REGULAR / FOLLOWING AM
		STAFF OR DIVISION: FD P230	TELEPHONE: 30686 — DATE: 1/5/68	TODAY SURE / MAILGRAM

TYPES OF DELIVERIES ARE DESCRIBED IN THE FORD TELEPHONE DIRECTORY FOR SOUTHEASTERN MICHIGAN

To: Mr. Jack Muir
Wixom Assembly Plant
Teletype Code LG

This will confirm our phone conversation requesting you release five (5) 428 Cobra Jet Mustangs to be picked up at Wixom Assembly Plant as follows:

VEHICLE ID#

Holman & Moody Truck (pick up 1/6/68) — 8F02R135050 ✓ JON IAC
8F02R135052
8F02R135053
8F02R135054

Ship to: Holman & Moody-Stroppe
2180 Temple Avenue
Long Beach, Calif.

Mr. Bill Jensen (Pick up 1 /5/68) — 8F02R135032
Ford Division - Performance & Economy Sect.

Ship to: Pick up by B. Jensen

C. R. Foulger, Stock Vehicles

Landy's Dodge Dart in the class final. Harvey, Platt, and Joniec handled the four-speed cars in the SS/E class. Platt did double-duty by driving the Tasca Ford entry in C/SA, but he would miss the call for the C/SA class final and was disqualified.

On race day for SS/E I think I had to run against two cars before I would face Hubert Platt for the trophy run. Both of us could run well under the record, which put us in danger of ruining our handicap in the Eliminator if we didn't take it easy.

Chuck Foulger had worked out a strategy to protect the handicap in red-light situations. Normally if you left first, there was no way to know if the car you were racing had red-lighted behind you. Chuck asked Al Turner (the Mercury drag coordinator) to station himself three-quarters of the way down the strip and wave at us if our opponent red-lighted. This way we would be sure to slow down and not breakout and/or lower the record and ruin our handicap.

Super Stock racing was/is based on cutting a great light and running as close to your record as possible. Thus Super Stock racing was full of red lights. Cutting a great light was so important that if you were not red lighting once in a while, you were not trying hard enough.

I was always a heads-up racer and this type of racing was not my favorite. I just didn't like it. However, I understood it well from the last two years of NASCAR racing, where I had to run first

Why is this young man smiling? Well, besides getting to meet Miss Winternationals, Al Joniec has just accomplished several tasks. First, he debuted the 428 Cobra Jet Mustang on behalf of Ford in spectacular fashion. Second, he not only took home the SS/E class victory but became the first Ford driver to win the Super/Stock Eliminator title at an NHRA event. And last, but certainly not least, Al broke the jinx of always being the bridesmaid at National events.

Al receives congratulations for his historic victory from NHRA President Wally Parks for the first of many wins for Ford cars powered by the new 428 Cobra Jet engine.

heads-up and then handicapped in the multi-class eliminator races. So I played the game, knowing that my chances of winning were just as good as anyone's because I could easily dip below the record, so it was all about cutting a great light.

Hubert and I and our two team cars were using an even start and no one was sure what to do. As we were warming our engines and putting our seat belts on to get ready to race, we still had no plan. I still remember Chuck Foulger's frantic-looking face as he opened my door at the very last minute and yelled, "Whoever is ahead by the end of 3rd

Upon his triumphant return following his Winternationals win, Rice and Holman Ford, Al's sponsor, displayed the car and trophy in their Pennsauke showroom. Win on Sunday, sell on Monday at its best. Jardine headers, the company contracted by Ford to fabricate the exhaust headers for the factory Cobra Jet team cars, also took advantage of Al's win in their ads.

Following the Winternationals, Al's Cobra Jet Mustang received stripes of blue Velvetex finish like that used on *Hairy One*, and the Mustang became known as *Hairy Two.* After breaking the 428 Cobra Jet, Al installed an all-out 427 Tunnel Port engine for match racing and participation in an all new heads-up Super/Stock series against the likes of Bill Jenkins and Ronnie Sox. This concept would give birth to what later became known as Pro/Stock.

The most famous of all Cobra Jet Mustangs would eventually be sold off to make room for new racing ventures in the Joniec stable. But, in a fashion similar to that of his 1963 Galaxie, the old race car would continue its winning ways after finding its way into the hands of long-time Ford man Dick Estevez of Maryland.

FOR SALE

Al Joniec's Hairy Too SS/E Mustang

Winner of 1968 Winternationals. Ready to race. Newly rebuilt engine. $3,800. Call (215) 672-2750 days or 331-1793 nights.

gear wins," so that we would be sure to back off after that to avoid going below the record.

As fate would have it, Hubert red lighted but I didn't/couldn't see it, as he was out on me about 1-1/2 to 2 cars off the line. Boy, did that get my attention! I said to myself, "Holy shit," either he was sandbagging me and had his shit together, or I was late on the lights. Whatever it was, he was out there in front. I went after him hard and I didn't pace myself this time as I really went through the gears. At the top of second I could see I was closing on him, and by the end of third gear I was about to pass him. When I put it in 4th I was ahead of him. As I leaned back and caught my breath, I saw this crazy man in the corner of my eye, standing on the return road fence and waving his hands like a nut bird. It was Al Turner, and I said to myself, "Hubert red lighted!!! *That* was why he was so far ahead of me off the line!" (I assumed Chuck had told him that I could outrun him, and that had pushed him to red-light not just once but also later in the eliminator). I backed off and let him go; if I had continued at the pace I was going, I would have buried the record and could never have won Super Stock Eliminator. I had won the first phase of the race, the SS/E class and looked forward to the Super Stock Eliminator runoff.

I really didn't understand how important this race was for Chuck Foulger and Ford, and I'm glad I didn't—it would have made me nervous, too. I did everything I was supposed to, as the car looked good and ran great, so I was totally relaxed. I also was used to driving my fast *Batcar* (now the *Hairy One*) SOHC altered-wheelbase Mustang, which made the Cobra Jet feel like I was driving in slow motion, and I could tell by the seat of my pants what was happening in the car better than most. Bottom line, I had the performance, I just needed consistency to win. That included cutting great lights. For Super Stock Eliminator it was like racing the whole company called Mopar; except for Hubert, it seemed like I had to race nothing but Mopars. Apparently the Mopar teams knew the Cobra Jets could easily run below the record. I assumed they had a strategy to force the CJ out by becoming martyrs and red lighting to force us to run below the record and therefore lose. There were still four Mopars left for me alone to run as one after the other continued to red light. The first car I remember running and beating was Don Grotheer's Mopar. In the second race I beat Tommy Crutchfield's Mopar, and in the third race I again faced Hubert Platt. Hubert was in the eliminator because he qualified low against the record, but he red-lighted against me yet again.

Fourth to go down was Dick Landy's car, driven by Butch Leal (I loved beating him as revenge for the '64 Nationals), who ran off an 11.27-second index. Little did he know, I probably could have outrun him in a heads-up race. I did need Al Turner's help calling my opponents' red lights on some on some runs, but most times I could stay ahead of them to the finish line while backing off and running out even though they red lighted—all due to cutting

After a long racing career, the Winternationals-winning Cobra Jet Mustang was found and restored back to its former glory. It was one of the featured cars at celebrations of the 40th anniversary of the introduction of the Cobra Jet. Here Al is seen making a pass in his old mount at South Mountain Dragway as part of the Carlisle All-Ford Nationals in 2008. (Bob Smith photo)

great lights. I remember the young H&M crew guys who were carrying ice, water, tire pressure gauges, wrenches, and so on, as well as towels for the drivers between runs for all of the other team cars. I watched these really great guys working their asses off, doing whatever it took to help, and it was really great to see.

By the semifinals there were no other Ford cars but me. So these crew guys came over to me and asked, with their can do/whatever-it-takes attitude, if I would like any help? But between every round all I had done was simply open my hood halfway to let the hot air out. I didn't need to go faster, and I didn't want to do anything else as I wanted to maintain consistency by doing the same thing over and over. I even limited myself to one small burnout each round to clean my tires off before I ran, so as to save the equipment. They wanted to help me so badly but I had to tell them, "Thanks, but no thanks," as they stood there in disbelief, sweating, with all of their stuff at the ready.

The final trophy run for Super Stock Eliminator was against Dave Wren, yet another Mopar, and I won! Then the celebrations began. I was immediately stopped at the scales to make sure the car was over the minimum allowable weight. Then I was chaperoned by a NHRA official, who rode back with me to the NHRA winner circle and then the thorough post-race tech inspection. The H&M guys that wanted to help before now had something to do, and they did most of the work. What the hell, they had put it together!

While I was in the winner's circle, here came a small motor scooter with a wide body on it. Lo and behold, it was Bill Jenkins. He was smoking his trademark cigar as he walked over to congratulate me and shake my hand. It had been only three years since he won the same event in the car I was supposed to be driving. He said, "You know, this is the very first Super Stock Eliminator that Ford ever won."

I started to realize this victory was a real home run for me and Ford. They had introduced the brand new Cobra Jet Mustang, won the SS/E class, and won the Super Stock Eliminator for the first time. Others were telling me it was historic and I was just starting to understand why."

WHAT ARE THE ODDS?

Four decades after Al's original Winternationals win, Ford decided to once again produce a limited edition (50 units), drag racing-only Cobra Jet Mustang in 2008. The first ten of these special vehicles were purchased by Brent Hajek of Ames, Oklahoma, a renowned collector of historic Ford race vehicles. In true blue fashion Brent had several of his cars prepared for competition at the NHRA Winternationals and lettered to match the original 1968 team cars. One of the new Cobra Jets was painted like the Rice and Holman Ford entry of Al Joniec, and was to be driven by long-time Ford racer John Calvert. The odds were long that a Cobra Jet Mustang would win the Super Stock Eliminator title forty years hence, and even longer that the one lettered as Al's original car would end up in the winner's circle. But that is exactly what happened, as John Calvert proved that lightning can strike twice.

Brent had also seen to it that Al, along with original 1968 drag team members Gas Ronda and Hubert Platt, were on hand for this historic occasion. Al found himself in the limelight once again thanks to the Hajek-Calvert team victory and, needless to say, Ford gained incalculable advertising points with the new Cobra Jet Mustang.

My crew and I had a celebration dinner the next day at a high-end restaurant because we had been eating the equivalent of roadkill out of a brown

Al Joniec poses with the new Cobra Jet Mustang at the 2008 NHRA Winternationals in Pomona, CA. Each of owner Brent Hajek's entries was lettered in the livery of an original Ford Drag Team 1968 Cobra Jet Mustang. This particular car, with long-time racer John Calvert driving, duplicated Al's feat by winning Super Stock Eliminator 40 years after the first Cobra Jet triumph.

The 428 Cobra Jet engine in the former Al Joniec Mustang has been faithfully and accurately restored to the same configuration he and the other Ford Team members used at the 1968 Winternationals.

bag all week. I told Bill Boyce and Ralph Hart to order anything they wanted. When I saw they had flaming cherries jubilee on the menu, I ordered that for everyone for dessert and we all ate like kings. Bill still talks about the flaming cherries jubilee even today.

After all of the partying and celebrating was over, we still had to drive at least four days or more to get home. We were not about to banzai home like we did coming out; I planned stay at a motel at least a couple nights. We were all tired, especially coming off the adrenaline high of winning the race, as we started the long ride home.

The first night Ralph was driving in the eastern part of the Rocky Mountains. I woke up because I had to pee and told him to pull over. As I stood alongside the truck peeing, with bleary eyes I looked back at the Cobra Jet on the trailer and noticed it was on a slant. When I finished relieving myself I walked back for a better look-see, and then I really woke up. The wheel was missing from the left side of our single axle trailer! We had been dragging the car and trailer on only one wheel. The trailer fender was bent way up, and it even made tire marks on the side of the car. I yelled to the guys to come back and see this disaster. In a few more miles we could have easily lost the car and trailer on the mountain road. My pee break saved the day.

Good old Ralph had dragged the trailer far enough to have completely wore off the spring U-bolts and half way into the nuts. There had been at least 3 inches of U -bolt sticking out of the nuts on the other side. We must have been creating quite a light show, as we had to have been sparking for miles.

I said, "Christ, Ralph couldn't you feel something was wrong?" He said he felt something but he thought it was just a rough road (we were on a beautiful interstate highway). Poor Ralph was just as beat as the rest of us and was driving in a tired stupor, just trying to stay awake.

I didn't know anything else to do but start walking back to search for the wheel. We walked for quite a long time before we found it. Apparently the axle nut came off the spindle and the entire brake drum wheel assembly took off. I still can't believe we found it at night. It didn't look too damaged, as the outside wheel bearing was retained by a C-clip. We rolled it back to the truck and checked for further damage, but it looked like all we had to do was jack up the trailer, put the wheel back on, and find a new nut. We were lucky to find a nut that fit in our spare nut and bolt box. We drove home and it didn't give us any more trouble, but we did check it often along the way.

In retrospect I had been just as careless as Ralph—on the way to the race, in the same area of the high Rockies and after about 40 hours of straight driving, I nodded off and drove off the road. I hit a road sign mounted on an angle-iron pole at around 65 mph. The sign pole bent over like a rubber band and the flat sign slapped the hood so hard it put a big dent in it. It caused the hood latch to open and the hood flew up. Bill and I were sleeping (I was supposed to be driving) until we were awakened by this big-time slap noise. We both woke right up, but we couldn't see out with the hood covering the windshield. Bill started screaming, thinking we were going to go off the side of the mountain. Ralph was in the back seat sleeping and didn't wake up until Bill screamed, but at least he didn't yell too. I was just waking up too, as I like to shit my pants trying to figure it all out, not being able to see out of the windshield and not knowing what was happening. I instinctively clamped down on Bill's shoulder to keep him from doing something stupid until he completely woke up. He says he still has the mark on his shoulder where I grabbed him. Luckily we were off the road on a safe, flat section and nothing happened. We straightened the large dent in the hood as best we could, closed it like nothing happened, and kept driving. By this time all of us were wide awake and silently thinking to ourselves how lucky we were to have escaped the grim reaper.

The funny thing was that we had towed the *Hairy One* 3,000 miles to the Winternationals and even checked it through tech inspection, but we never ran it in the race because of our focus on the Cobra Jet. However, we did take it out to Lions Dragway while we were in California, where the fur paint job was a big hit with the fans who had never seen such a thing. The car ran well, in the low nines at close to 150 mph.

In the Wake of the Winternationals

Back home I settled down and got ready for that year's NASCAR series. We painted the Cobra Jet with a racing stripe down the middle and the name *Hairy Too* in blue Velvetex fur paint to match *Hairy One*. I do recall having one match race with the car as a SS/E against a Chevy SS/D car. The Chevy was in a higher class but we ran comparable times. I was also a member of the Holley Carb Team back then and they sponsored me with the carburetors of my choice. I had obtained the all-new 3-barrel carb to try, as well as some others. I put the 3-barrel carb on while at the match race just to try it and damned if the car didn't go faster, so I left it on.

When I raced the Chevy I easily beat him, but it was a best of three. Don't you know, he started pissing and moaning about the carb (I never figured out how he knew I even put it on). I pointed out that this was a match race, so there was no tech inspection, plus he was in a higher class than me—I should have been the one complaining, but I wasn't. I went on to win that race, but I never ran the car as a legal Super Stocker again, except for the Summer Nationals in Englishtown, where the engine dropped a hollow-stem exhaust valve during time trials. I removed the S/S engine from the car and put it in the corner. I then put together a 427 Tunnel Port engine to go match racing and also to run in the 3,200 lb NASCAR Ultra Stock class. I match raced against Bill Jenkins a lot that year and I never remember losing to him.

During the 1968 season Al and his 427 Tunnel Port-powered Mustang participated in the fledgling Pro Super Stock series against the likes of Bill Jenkins (427 Camaro) and Ronnie Sox (426 Hemi Barracuda), among others. These matchups proved very successful with racing fans and very likely gave rise to the NHRA Pro/Stock class.

I received a call from the newest Ford Drag Coordinator, Emil Loeffler, and he invited me to come to the Dearborn Office. Chuck Foulger had worked out a deal to get a new Ford dealership and was no longer working directly for Ford; Emil was his replacement. I went to Ford headquarters and we talked about his vision of the future of Ford drag racing. I was excited; I knew that Ford wanted to focus on super stock cars that Ford could sell and I had just won the biggest such race so far. I assumed I would be included somehow.

Part of the re-emphasis on showroom cars you could take drag racing was to introduce a new Ford "High Performance" parts book. Emil had a copy for me to review and was apparently looking for compliments in the company of several of his staff, who were also waiting for my feedback. The book looked nice and was professionally put together like a typical parts book. The book had all the common high-performance parts, but the good stuff—the "hard to get" parts that I thought racers would like most—were missing. I asked him, where were the special cams we were using, where were the lightweight valves, the dual-disk clutches, lightweight forged pistons, and so on. I blurted out that I thought you would want to help the Ford guys, just like Chrysler and GM did. They would list all kinds of trick parts and make them available for everyone to buy direct.

Apparently embarrassed, he said they had to save that kind of stuff for our team. I learned later that he came from the truck division and didn't know anything about what was going on regarding racing. I guess I could have been more tactful, but he did ask my opinion. I really didn't understand the politics of life back then, so I spoke from my heart. I consider myself a no-BS straight talker, even today. If you ask me for my opinion I will tell you the truth (just ask my wife). I always believed in knowing the truth. You don't have to like it, but I always felt that you should know it. It was some time later that I learned he had asked me to come to Dearborn by myself to size me up for the upcoming professional traveling Ford Drag Team performance seminar/clinic concept that was a part of the new Ford drag racing image. Apparently I failed the test.

Ford Motor Company followed Chrysler's lead in 1969 by forming two Drag Teams that not only raced specially prepared and painted Super Stock Ford products, but that also traveled the country to put on high-performance clinics at Ford dealerships. The team of Ed Terry and Dick Wood comprised the West Coast team while Hubert Platt and Randy Payne were the East Coast team. The western team cars and transporter were painted white with blue lettering and striping; the eastern team cars were blue with white lettering and stripes. The day the East Coast team of Platt and Payne held their clinic at our local Ford dealer, I recall the traffic was backed up for blocks from folks trying to get in.

I went back to racing both of my cars but it was not as easy as it had been. I didn't realize at the time that the workload was starting to take its toll on me. I did win my fair share of races and had an OK season, but I was getting burned out (I had been going at it for close to 10 years of 24/7, grinding work). I was doing everything myself and there were simply not enough hours in the day. I also never learned to hire help and/or delegate some of my workload. The bottom line was I didn't have the money and/or resources to do it right. The endless travel to get to races all across the country was eating me up even more.

At the end of the year it was time to sell my cars so I could get new ones for next year. I took the Tunnel Port engine out of *Hairy Too,* reinstalled the now-rebuilt Cobra Jet engine, and sold it locally for $3,500 and thought I got a good deal (little did I dream of how much it sold for years later). I sold the *Hairy One* to Drake Viscome of Carmel Ford in Carmel, New York for some money, a used car for my father, and a bulldog. I had fallen in love with Drake's bulldog and he offered me a pup from the next litter as part of the deal.

That bulldog was one of the best deals I ever made—Butch made every day a fun day for me. He was such a character, but he couldn't stand exhaust header noise and went crazy whenever I started a race car. He would squeeze himself under the low race cars and practically pick them up as he tried to bite the header pipes. He'd come up from under the car with burnt whiskers and a black face.

One day an open-pipe motorcycle stopped at the drug store directly across the street. The rider parked his old Triumph while Butch watched from the front step. I knew something was going to happen when I saw Butch start licking his chops (a sure sign he was on to something). When the guy kickstarted his bike, Butch burnt nails running after him. Butch bit and grabbed the one of the pipes and shook it. The poor guy didn't understand what was happening with all of the noise and the whole bike shook. He finally

Here is the famous Butch the Bulldog, who came into Al Joniec's life as part of the trade deal with Drake Viscome involving the *Batcar.* Al counts Butch as one of the best deals he ever made, as the dog provided years of laughs.

leaped off the bike and ran as the pipe came off and Butch proudly ran back to show me the pipe. The poor guy didn't know whether to shit or go blind as he looked on in awe. I recovered the pipe and put it back on the bike for him, and I put Butch in the store and closed the door while the rider started the bike once again and drove off into the sunset.

MORE BUTCH

One of my favorite stories of Butch, as told in typical Al Joniec fashion, is the seemingly bigger-than-life canine's relationship with the U.S. Postal Service. While it is not uncommon for a dog to view the local mail carrier as a usurper then they are making their deliveries, Butch took it to a whole new level.

What began with Butch shredding the mail as it was pushed through the slot in the front door of the shop finally escalated to an incident where Al received a call from the local police requesting his presence at the shop. It seemed that Butch had gone through the glass door leading to Al's shop in his efforts to protect the premises from the mail carrier. Upon his arrival on the scene, Al found Butch sitting on the step, shattered glass all around, looking quite pleased that he had run off the mailman. Luckily there were no injuries on either side of the incident, but the police did request that Al do something to secure the shop against future incidents. The plain door glass was replaced with a pane that was wire-reinforced to keep Butch from breaking through.

Imagine Al's shock when he received yet another call from the police, summoning him to the shop to corral Butch after yet another attack on the mailman. Al describes the scene: "Here is Butch sitting on the porch looking very proud and behind him are the smashed remnants of the window, with the outline of a Bulldog's head clearly visible in the wire mesh."

CHAPTER SEVEN

MAJOR CHANGES ALL AROUND

FOR ALL INTENTS AND PURPOSES the 1969 season began where the previous year left off for Al, with yet another trip to the NHRA Winternationals in Pomona, California.

I got a call from Ford and they said there would be a brand new 1969 Cobra Jet fully prepared for me to run at this year's Winternationals. They said it would be completely lettered and race-ready and they would even supply a truck as well. All I had to do was bring my helmet and fly out. I said, "Hell yeah, I'll be there," and true to their word, the car was there waiting for me. The only thing I was concerned about was if it would be competitive, so I went to a local track during the week to test. The place was packed with tuners and local grudge racers preparing for the Winternationals. So here I was in California with a packed house, and who do you think pulls up to race against me? It was my neighbor from Pennsylvania, Bill Jenkins. His winning 427 Camaro car was in a higher class than me (I think SS/D), since the Cobra Jet Mustang was now in SS/I. During testing I didn't have to worry about how fast I was going and just wanted to see how fast the car was. I just beat the snot out of him as I easily pulled away in 4th gear. Unlike the 1968 car, this car was great off the line (most likely due to better tires). Apparently Stroppe's guys figured out how to make good power over the last year as the car could run 11 flat, well below the class record. The car felt really fast to me and absolutely flew for what it was.

I almost felt guilty, as I didn't have to do any work on the car whatsoever, other than put some weight in it even though the '69 was a bigger car than the '68. At the race Bill Jenkins came over and asked what the hell I did to the car, as he never got beat so bad as he had the other night.

I really felt bad on race day as I ran out and lost right away. I would like to blame the car for being so fast, but it was my fault. The car I was racing against was doing what all S/S handicap racers do: they go as fast as they can against the handicap, but not any faster than they have to. I was out of practice, most likely didn't cut a great light, and simply didn't want to lose. I didn't pay close attention to the handicap. It even caught me by surprise as I thought I would have done much better and not fallen into the trap of going too fast. At any rate, this quickly ended one of the most pleasant times I ever had racing as a celebrity guest in La La Land. It's ironic that I only saw this car for a few days and never saw it again. I have no idea what became of this great car.

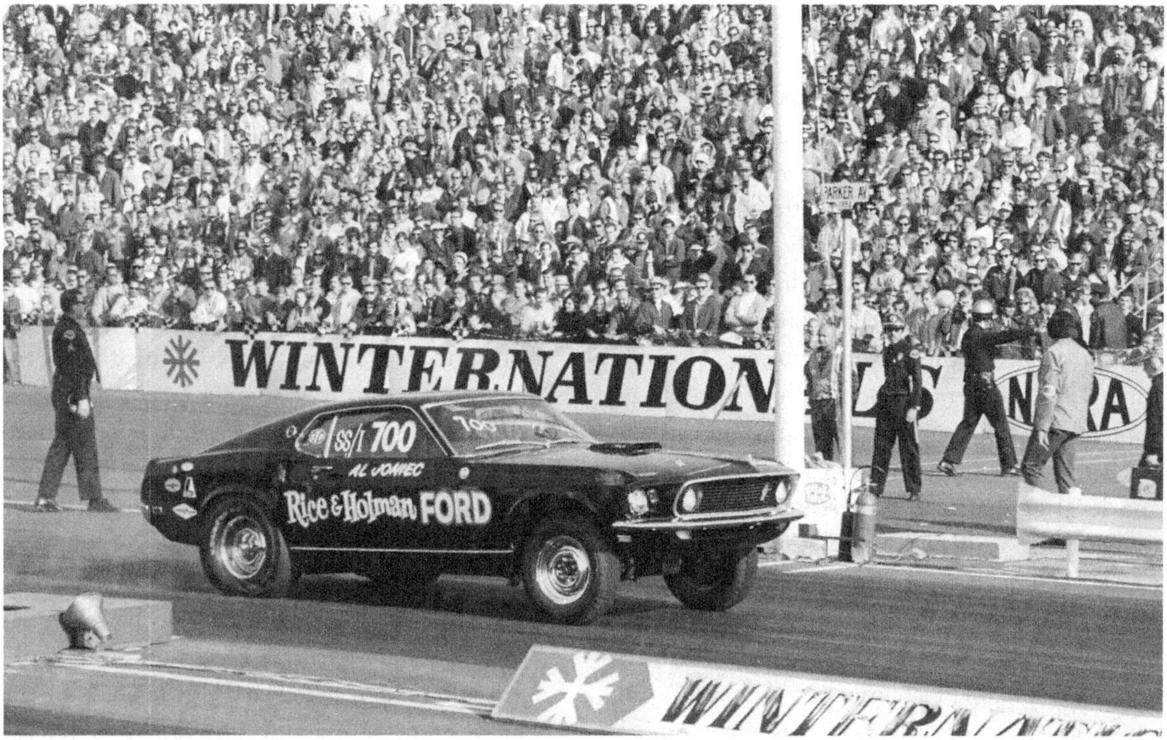

Ford had H-M Stroppe prepare a 1969 Cobra Jet Mustang, painted in Rice and Holman livery, for Al to compete with at the 1969 Winternationals. All that was required of Al was to fly in and drive. In pre-race testing the car proved to be superior to the ‘68 Mustang in performance and things looked good for a repeat victory. Unfortunately for Al the faster car proved to be his downfall as he eliminated himself by running quicker than his class index in the first round of competition. Al flew home and the car remained in California.

Enter the Boss 429

I asked Al if he ordered a '69 Boss 429 Mustang, believing it to be the next great performance idea from Ford, and with developing that engine in mind. Al replied that in his estimation the Boss 429 engine was a slug; the sole reason he bought the car was the fact that Kar Kraft Engineering had modified the dimensions of the engine compartment to accommodate the Boss 429 power plant, which would in turn allow sufficient room for easy installation of the SOHC 427. On June 6, 1969, after paying the princely sum of $3,994.98 and other fees, which brought the total to $4,154.78 for the new Boss Mustang (KK 1720) at Rice & Holman Ford, it was driven just a few miles to Al's shop where its transformation into a race car began.

I ordered a brand new 4-speed Boss 429 from Rice & Holman for the 1969 season. I had measured the engine compartment of one previously to make sure it would fit a Cammer, and it did.

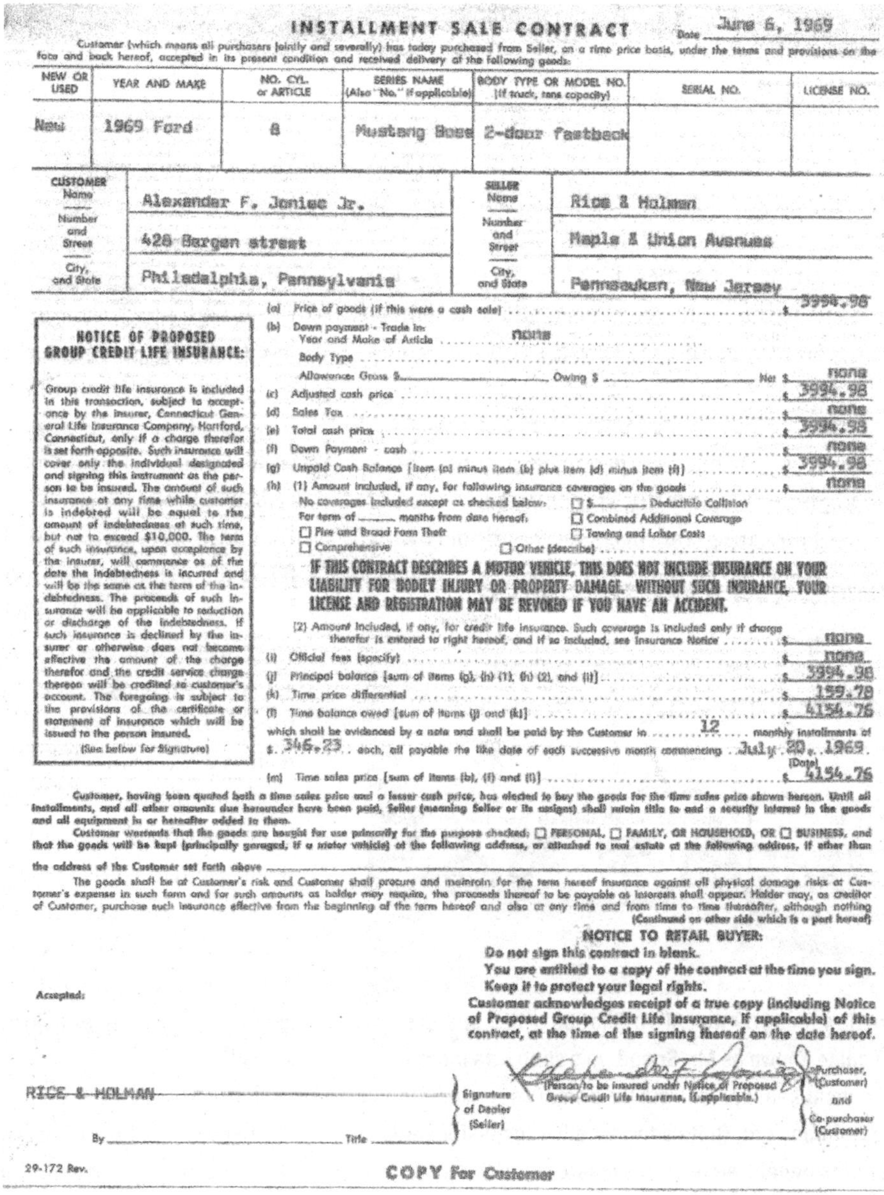

INSTALLMENT SALE CONTRACT Date June 6, 1969

Customer (which means all purchasers jointly and severally) has today purchased from Seller, on a time price basis, under the terms and provisions on the face and back hereof, accepted in its present condition and received delivery of the following goods:

NEW OR USED	YEAR AND MAKE	NO. CYL. or ARTICLE	SERIES NAME (Also "No." if applicable)	BODY TYPE OR MODEL NO. (If truck, tons capacity)	SERIAL NO.	LICENSE NO.
New	1969 Ford	8	Mustang Boss	2-door fastback		

CUSTOMER		SELLER	
Name	Alexander F. Joniec Jr.	Name	Rice & Holman
Number and Street	428 Bergen street	Number and Street	Maple & Union Avenues
City, and State	Philadelphia, Pennsylvania	City, and State	Pennsauken, New Jersey

NOTICE OF PROPOSED GROUP CREDIT LIFE INSURANCE:

Group credit life insurance is included in this transaction, subject to acceptance by the insurer, Connecticut General Life Insurance Company, Hartford, Connecticut, only if a charge therefor is set forth opposite. Such insurance will cover only the individual designated and signing this instrument as the person to be insured. The amount of such insurance at any time while customer is indebted will be equal to the amount of indebtedness at such time, but not to exceed $10,000. The term of such insurance, upon acceptance by the insurer, will commence as of the date the indebtedness is incurred and will be the same as the term of the indebtedness. The proceeds of such insurance will be applicable to reduction or discharge of the indebtedness. If such insurance is declined by the insurer or otherwise does not become effective the amount of the charge therefor and the credit service charge thereon will be credited to customer's account. The foregoing is subject to the provisions of the certificate or statement of insurance which will be issued to the person insured.

(See below for Signature)

(a) Price of goods (if this were a cash sale) .... $ 3994.98

(b) Down payment - Trade In: Year and Make of Article .... none
Body Type ....
Allowance: Gross $ ________ Owing $ ________ Net $ none

(c) Adjusted cash price .... $ 3994.98

(d) Sales Tax .... $ none

(e) Total cash price .... $ 3994.98

(f) Down Payment - cash .... $ none

(g) Unpaid Cash Balance [item (a) minus item (b) plus item (d) minus item (f)] .... $ 3994.98

(h) (1) Amount included, if any, for following insurance coverages on the goods .... $ none
No coverages included except as checked below: ☐ $ ______ Deductible Collision
For term of ______ months from date hereof; ☐ Combined Additional Coverage
☐ Fire and Broad Form Theft ☐ Towing and Labor Costs
☐ Comprehensive ☐ Other (describe) ....

IF THIS CONTRACT DESCRIBES A MOTOR VEHICLE, THIS DOES NOT INCLUDE INSURANCE ON YOUR LIABILITY FOR BODILY INJURY OR PROPERTY DAMAGE. WITHOUT SUCH INSURANCE, YOUR LICENSE AND REGISTRATION MAY BE REVOKED IF YOU HAVE AN ACCIDENT.

(2) Amount Included, if any, for credit life insurance. Such coverage is included only if charge therefor is entered to right hereof, and if so included, see Insurance Notice .... $ none

(i) Official fees (specify) .... $ none

(j) Principal balance [sum of items (g), (h) (1), (h) (2), and (i)] .... $ 3994.98

(k) Time price differential .... $ 159.78

(l) Time balance owed [sum of items (j) and (k)] .... $ 4154.76

which shall be evidenced by a note and shall be paid by the Customer in ..... 12 ..... monthly instalments of $ 346.23 each, all payable the like date of each successive month commencing July 20, 1969. (Date)

(m) Time sales price [sum of items (b), (f) and (l)] .... $ 4154.76

Customer, having been quoted both a time sales price and a lesser cash price, has elected to buy the goods for the time sales price shown hereon. Until all installments, and all other amounts due hereunder have been paid, Seller (meaning Seller or its assigns) shall retain title to and a security interest in the goods and all equipment in or hereafter added to them.

Customer warrants that the goods are bought for use primarily for the purpose checked: ☐ PERSONAL, ☐ FAMILY, OR HOUSEHOLD, OR ☐ BUSINESS, and that the goods will be kept (principally garaged, if a motor vehicle) at the following address, or attached to real estate at the following address, if other than the address of the Customer set forth above ________

The goods shall be at Customer's risk and Customer shall procure and maintain for the term hereof insurance against all physical damage risks at Customer's expense in such form and for such amounts as holder may require, the proceeds thereof to be payable as interests shall appear. Holder may, as creditor of Customer, purchase such insurance effective from the beginning of the term hereof and also at any time and from time to time thereafter, although nothing

(Continued on other side which is a part hereof)

NOTICE TO RETAIL BUYER:

Do not sign this contract in blank.
You are entitled to a copy of the contract at the time you sign.
Keep it to protect your legal rights.

Customer acknowledges receipt of a true copy (including Notice of Proposed Group Credit Life Insurance, if applicable) of this contract, at the time of the signing thereof on the date hereof.

Accepted:

RICE & HOLMAN ________ Signature of Dealer (Seller)

By ________ Title ________

________ (Person to be insured under Notice of Proposed Group Credit Life Insurance, if applicable.) Purchaser, (Customer) and ________ Co-purchaser (Customer)

29-172 Rev.

COPY For Customer

Al purchased a Boss 429 Mustang from his sponsoring dealership, Rice and Holman Ford. The new car was then driven just a few miles to Al's shop, where its transformation into a race car began. This copy of the original sales agreement indicates Al paid $3994.98 for the car and took delivery on June 6, 1969.

Before the new Mustang had even been lettered, Al had it out at Atco Dragway to get it dialed in and test its feasibility for competition in the classes he had chosen to run in 1969. While the fender graphics on the car still read Boss 429, power is being supplied by the SOHC 427.

I drove the car from Rice & Holman to my shop and proceeded to take the Boss 429 engine out and put it in the corner. I also took the bucket seats out and had some brackets made to use them as a decorative seat for my house. I still have them today, almost 50 years later.

I then had my good friend Al Rowe make me a roll bar and install my pre-made traction bars. The Cammer fit perfectly as planned and the big task of fabricating the headers was done by JR headers without a hitch. I got a fiberglass hood and put a large hood scoop on it right over the two 4-bbl carbs. A fiberglass bumper was installed in the front only. I took any and all accessories out, like the radio and heater (plus controls), rugs, and rear seat, and so on to lighten it up as much as possible. I was sure to get real magnesium wheels (not aluminum), again to reduce weight but to also reduce rotating mass. I put the biggest front tire I could find in the front. The idea was to have as big a diameter tire as I possible so as to keep the first stage light blocked while still rolling forward without uncovering it to reduce red lights.

Also new for 1969 was Al's affiliation with a speed shop franchise known as Mr. Speed. Al's shop was located in Pennsauken, not far from his primary sponsor, Rice and Holman Ford. Quite a few locally successful Ford products benefitted from Al's masterful touch behind these doors.

Another trick was to take off the rear deck lid torsion bar. This shifted all the weight from the front of the trunk lid and splits half the weight to the rear. It's a small trick, but it really helps gain traction. I learned this trick back in 1962 when we had to use true street tires and the rear deck lids were very long and very heavy. I believe I picked up a full tenth of a second back then with this trick, and I did the same trick to every car since. Obviously I put the battery in the trunk as well.

I made sure to take out the standard toe-in of the front-end alignment (the toe-in helps stock cars go straight and not wander going down the highway) and make it as close to zero toe as possible in

Another entry into the Joniec stable of racecars for 1969 was his one and only flip-top Funny Car, based on a '69 Mustang body. The car, which was powered by a fuel-injected, gas-burning SOHC 427, proved too heavy to be competitive in the rapidly changing NASCAR drag race series competition and it was soon sold off.

The Rice and Holman Ford tow rig with Al's Boss 429 and Funny Car Mustangs parked in front of the dealership.

Al blasts off from the starting line at Atco Dragway in New Jersey with his SOHC 427-powered 1969 Mustang.

In an early Pro/Stock match race (called Pro/Super Stock at the time), Al faces the formidable factory 426 Hemi-powered Barracuda of Jack *Mr 5 and 50* Werst.

The power of the Al Joniec prepared SOHC 427 is evident in this photo as he wrinkles the sidewalls of the slicks off the starting line.

all of my cars. Dragging your front tires sideways (even an eighth of an inch) slows you down.

Other changes included modifying the shifter hole for a short-throw Hurst shifter. A shorter drive shaft was required as well, so I had one fabricated. These changes gave me a fast, cheap, and easy way of getting into the 3,200-pound category that was going on at the time. I also match-raced Bill Jenkins (among others) who was my competitor in our 1968 wars. The fans loved to see Ford against Chevy. I liked racing Bill because I felt he stayed true to the super stock creed. He didn't use a mountain motor and/or use other cheating ways that others sometimes did.

I also felt that it was Bill who pushed the Pro Stock class with the political powers at NHRA the following year. After all, Ford wanted to end the crazy Funny Car stuff in order to sell cars from the showroom.

I was also still running in the now-much-less-popular NASCAR circuit; I believe that 1969 was the last year for them. NHRA had fully embraced the fuel Funny Car deal, which now had rules and had become very professional. Anybody that ever saw, smelled, and felt them run (they shook the grandstands and you could feel the power in your chest) had to be impressed. It was impossible for the NASCAR gasoline cars to compete with them. As a part of my sale of the *Hairy One* to Carmel Ford, they agreed to help me find a used flip-top Funny Car to use on the NASCAR circuit. There were several flip-top cars that I was competing

In this match-up Al faces off against Bill "Grumpy" Jenkins in his 427 Camaro. Al fondly relates how Jenkins was his first match-race victim with the 1969 Mustang.

When found the former Al Joniec '69 Mustang had been repainted and re-named *Ford Grabber.* At this point the car had not been raced in a number of years and was sitting beneath a tarp in the owner's driveway. (Spinelli family photo)

with on the circuit and I felt there was more to going fast then just having a strong engine. I got the call to look at the Art Malone car, which was a 1969 Mustang body with a professional chassis built by Woody Gilmore, but without an engine or transmission. The car looked great and I liked the quality of the chassis so I made the deal to buy it. After making the deal, as I put the car body back down on the chassis I couldn't believe how heavy it was. Because the car had no engine or transmission and I couldn't weigh it, I had to assume it was going to be lighter than my old *Batcar.* That would prove to be a big mistake.

I made the conversion, installed the injected Cammer, replaced the rear housing and brakes, had the car painted, and got it ready to race. I thought the car looked great, but after all of the hard work the car weighed about the same as the *Batcar* (2,675-plus pounds) mostly because of the incredibly heavy body. Because I was in a 2,600 lb class, the goal was to get the 75 pounds out of the car, which should have been worth around 5 to 7 hundredths of a second of ET. The end result was that the car didn't go any faster than the *Batcar*. Aside from the fact that I could hardly pick up the body by myself, the disappointing results turned me off so badly that I only ran the car a few times before I parked it.

I later won my fair share of the 3,200 lb class and won most of my match races with the Boss 429 car. I also won some but not as many as before in the 2,600 lb class, but I believe the NASCAR circuit permanently ended at the end of the year. During that time frame I got talked into being a figurehead for a new speed shop franchise named Mr. Speed. There was already a company named Mr. Gasket, so some genius thought it was a good name. The franchise used an "S" shield logo, just like Superman. I wasn't intending to exploit superheroes, it just worked out that way.

When Al finished racing the 1969 Mustang, the SOHC 427 was pulled out and replaced with the original Boss 429. When purchased by the Spinelli family, the car still had Boss 429 power. (Spinelli family photo)

While Al's memories of his Mr. Speed franchise business are not the fondest, there are a great number of area racers who credit the preparation done to their cars at Al's facility with making them winners on the track during that era.

Al eventually sold the Boss 429 car as well. Like most old race cars it faded from memory for many years. Then in 1982 Jerry Spinelli spotted the unmistakable outline of a 1969 Mustang under a cover sitting on a trailer outside of a modest home in Marcus Hook, Pennsylvania. He stopped to speak with the homeowner and was able to learn that the car was indeed a Boss 429 Mustang and that it had been raced until a loss of interest and the responsibilities of life caused the owner to park it. Jerry was able to negotiate a deal on the car and brought it home. Eventually the car was sold to his brother Joe, who set out to locate the old race car's missing title. Joe's investigation led him to a man named Al Guglielmo, the car's second owner, who had retained custody of the title when a deal he made to sell the car in return for an F-250 pickup truck and $1,000 cash was not satisfied by the buyer, who failed to pay the cash portion of the deal. It was at this time that Joe discovered the car's original owner

continued on page 114

Joe Spinelli undertook a totally accurate restoration of the former Joniec mount, with insights and assistance from Al himself. When completed, the car would be better than new. (Spinelli family photo)

The restored Mustang was unveiled to Al Joniec at the York U.S. 30 reunion several years ago. A smiling Joe Spinelli (owner) looks on as a beaming Al Joniec turns the key to fire the engine in his former racer. Also on hand at the same event were the restored *Batcar* and Al's '68 Winternationals-winning Cobra Jet Mustang.

The completed Al Joniec Rice and Holman Ford/Mr. Speed Boss 429 Mustang is nothing short of spectacular in its appearance and adherence to period-correct accuracy.

Lloyd Wolfe took this candid photo of Al in the '69 Boss 429 waiting in the staging lanes during the 1969 season. From the X designation on the window of the adjacent car, it appears that in this case the class was Experimental Super/Stock, one of several terms used before Pro/Stock became the norm.

had been none other than Al Joniec. Realizing that he was in possession of a piece of Ford racing history, Joe reached out to Al. Over the next several years, while the car was restored, Al was most helpful with photos and information to aid in the process.

FORD PULLS THE RUG OUT FROM UNDER RACERS

Surprisingly enough, around the end of the 1969 season I got a call from Ford instructing me to plan on picking up my Ford Drag Team car and truck in approximately 90 days. I stopped all of my other activities and canceled anything that would conflict with the race and travel schedule while putting on seminars and so on. It was only about 60 days or so later that I got another call from Ford, informing me that Ford Motor Company had made a corporate decision to cease and desist all racing activities because of pending federal emission requirements. Ford would be applying all of the race engineers to the then-new emissions program. It was really over.

CHAPTER EIGHT

THE PRO/STOCK CLASS COMES INTO ITS OWN

NINETEEN SEVENTY started off with a really big bang for me, literally, but I'm getting ahead of myself. Although Ford and the other auto manufacturers were now officially out of racing, they stayed involved in the game in innovative ways via their huge advertising budgets. Nineteen seventy was the first official year for Pro Stock and the new hot setup for Ford that year would be a Maverick with a Cammer. I bought a new 6-cylinder Maverick at cost from Rice & Holman, drove it to my shop and took the engine out. I re-powered my 1960 Falcon Ranchero parts chaser car with the engine and transmission from the Maverick (it ran great with the 250 cid 6-cylinder). I then stripped the rest of the Maverick down to lighten it as much as possible while getting it ready for the conversion to the new Pro Stock class. I believe someone at Ford had arranged for Lakewood Engineering in Ohio to do the modifications the Maverick required, such as moving the original front spring towers outward to accommodate the width of the Cammer, as well as other things like clutch linkage, motor mounts, roll cage, and so on. My car was just one of many Ford and Mercury team cars as well as others non-team cars done by them that year. One thing that I didn't do a good job on was the rear wheelwell clearance. Even though I did open them up as much as possible and altered the rear wheel openings, the spring location restricted the maximum slick tire to 12 inches. The big stuff like the engine installation was easy because the exhaust system was already made by JR Headers. All I had to do was have the car painted. I was already biased toward a scallop-type paint job, even before the *Batcar* and going way back with the 1958 "Hoagie" Ford that had gold scallops. I got the master artist Ralph Hart to lay out the paint scheme and had Bob Cummons paint the car.

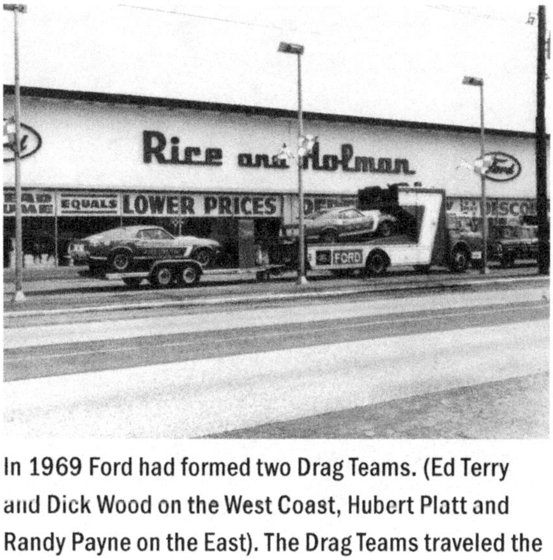

In 1969 Ford had formed two Drag Teams. (Ed Terry and Dick Wood on the West Coast, Hubert Platt and Randy Payne on the East). The Drag Teams traveled the nation, racing and putting on performance clinics at various Ford dealers.

I normally had my engine machine work done by Andalusia Speed & Machine (they did all the machine work on my engines at the time). One day while I was over there, a guy who just started a new business close by called Hot Rod Clutch Company came over and introduced himself. He asked if I would use his clutch as he felt it would be good for his business. I told him I run my engines as high as 9,500 rpm and I needed a strong clutch that could also release cleanly at these revs to shift gears. He said that would be no problem and that it would be the best clutch I would ever use. He seemed sincere and was just looking for a break to help his new business. I agreed and told him to send me one. I put my best engine in the car and installed the new 11-inch Hot Rod clutch. I finished setting up the suspension while I was waiting for the shortened drive shaft that would finish the car. When everything was done I went to the track to try it out. I took some slow passes down the track to check for vibrations and leaks and to get a feel for the handling and brakes. After I had gone over 100mph, I was satisfied everything was OK and decided to make a full pass. I did a burnout, went to the line, staged, and as the lights came down I revved it up to around 8,000 rpm. As I let out the cutch all hell broke loose! BOOM, the clutch exploded and like to cut the bellhousing in half. It cut the steering shaft, brake lines and generally ripped the hell out of the car. It even blew out the back of the engine block. Luckily I still had my feet and I wasn't hurt. I was not going that fast, but I couldn't steer and/or stop as I ran off the track and into the ditch with my brand new car.

Boy, was I pissed. Here I was, trying to be a nice guy, and look what happened. I had so much work to do if I was going to make any race dates that I had planned (I would miss a lot of them because of it), and I just blew out the back of my best engine. When I took the car back to the shop, it was very apparent that it was the clutch that had blown, as the flywheel was in one piece but the pressure plate was shattered in hundreds of pieces.

I took several of the broken pieces to a metallurgist and had them analyzed to see if it was nodular iron, like most high-performance clutches

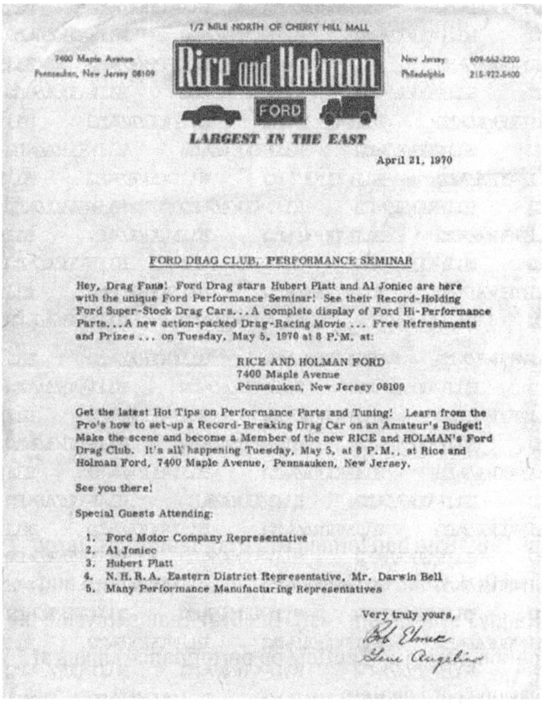
1/2 MILE NORTH OF CHERRY HILL MALL

7400 Maple Avenue
Pennsauken, New Jersey 08109

Rice and Holman FORD

New Jersey 609-662-2200
Philadelphia 215-922-5400

LARGEST IN THE EAST

April 21, 1970

FORD DRAG CLUB, PERFORMANCE SEMINAR

Hey, Drag Fans! Ford Drag stars Hubert Platt and Al Joniec are here with the unique Ford Performance Seminar! See their Record-Holding Ford Super-Stock Drag Cars...A complete display of Ford Hi-Performance Parts...A new action-packed Drag-Racing Movie ... Free Refreshments and Prizes ... on Tuesday, May 5, 1970 at 8 P.M. at:

RICE AND HOLMAN FORD
7400 Maple Avenue
Pennsauken, New Jersey 08109

Get the latest Hot Tips on Performance Parts and Tuning! Learn from the Pro's how to set-up a Record-Breaking Drag Car on an Amateur's Budget! Make the scene and become a Member of the new RICE and HOLMAN's Ford Drag Club. It's all happening Tuesday, May 5, at 8 P.M., at Rice and Holman Ford, 7400 Maple Avenue, Pennsauken, New Jersey.

See you there!

Special Guests Attending:

1. Ford Motor Company Representative
2. Al Joniec
3. Hubert Platt
4. N.H.R.A. Eastern District Representative, Mr. Darwin Bell
5. Many Performance Manufacturing Representatives

Very truly yours,

In April 1970, Rice and Holman Ford hosted the Drag Team, along with Al Joniec. Later that year Al was slated to receive his own Drag Team vehicles and transporter, but unfortunately a cut in motorsports funding at Ford ended the agreement before it happened.

In 1970 Al stepped up his racing game yet again with the purchase of a new compact Maverick. The car's 250 cubic inch 6 cylinder engine was soon replaced by an SOHC 427, and the car received a stunning paint job, complete with scallops. In this promo photo taken just after the car was completed, Al positioned himself strategically to hide the fact that the car did not yet have a hood scoop.

were supposed to be. The test showed it was made of common gray iron. With the test results in hand I went to the Hot Rod Clutch guy and told him, "You will give me $3,500 for damages or I'll sue your ass." Before this he denied that he did anything wrong, so he was either a fraud and/or incompetent, but in either case I was done with him. The $3,500 was my actual cost (I can't imagine what it would cost today), not counting labor. He did give me the money and I believe he went out of the racing clutch business right after that. I had to buy a new block, but Ford was now out of racing. It hurt not to be able to get things for free, plus the newer blocks were not as good as the old ones. I put the engine back together, but for some reason it never ran as good as it did before. This was just before people started using block plates while honing their cylinder blocks. Today the common knowledge is that there can be a 25- to 40-horsepower difference between a good or bad hone of the cylinders because of the sealing of the piston rings.

I had two carbs with air horns sticking up out of the hood for my initial test and PR photos, as I didn't know what kind of intake manifold I would wind up with. I was looking to get some type of tunnel ram. I ended up with not one but two ugly aluminum hood scoops (more on that later) and one fiberglass hood scoop integrated into the hood. At this time there was a debate about rear versus front air entrance for such scoops. NASCAR racers were using cowl and/or rear entrance intakes very successfully, though I felt intuitively that the front entry was the way to go. I have to admit that Bill Jenkins heavily influenced me to start with the rear-entry approach, because he was successfully using the cowl-induction method. I first tried the aluminum hood scoop with the

The powerful SOHC 427 stuffed between the front fenders of the Rice and Holman Ford-sponsored Maverick had little trouble wrinkling the sidewalls of the slicks when Al let the clutch fly off the starting line.

rear entrance and a smaller space in the front and I experimented with moving the distance from the inlet to the windshield cowl area to find the best configuration. The ability to pop-rivet small pieces to the base scoop configuration permitted me to then find the best opening and then make a proper fiberglass one and paint it. Again, I wasn't sure what type of intake configuration I was going to be using at this point either.

I was running just out of the 9-second range and winning a lot with the local competition, but I needed to step it up if I wanted to run with the big dogs. The Chrysler Hemis were the cars to beat. Sox and Martin won the first two Indy Nationals in Pro Stock and even Bill Jenkins couldn't beat them.

I had done everything I knew how to do to the Cammer years ago. The only thing left for me to try was the tunnel ram intake manifold. But the more I looked for one, the more it was apparent that I would have to make one myself. I already had a fuel injector intake manifold and I was at that time sponsored by Holley carb (I believe they gave you up to 10 carburetors), so I got four 500-CFM 2-barrels, which gave me one carb for every two cylinders. My rationale at the time was 4 x 500 CFM equaled 2,000 CFM, which I felt would do the job.

I took it to the track to try it out. Though I didn't expect instant results I was not happy with the

Al's Maverick in action at Suffolk, Virginia and Atco, New Jersey during the 1970 season. The car would go on to wear several distinct hood scoop styles in its racing career. This one was yet to be painted.

One of Al's innovations in an attempt to gain more power from the SOHC 427 engine in his Maverick was to cut the Holley Dominator carburetors apart so that they could be mounted more directly above the intake ports. While it took some figuring out, the combination proved successful.

power I was experiencing while just driving the car up and down the track. Out of frustration I took the hood off to look at the carbs and engine while it was running. When I revved it up around 5,000 to 6,000 rpm I couldn't believe what I was seeing. A fountain of raw fuel was dancing above the carbs at least a foot high!

I had to try something. I thought that maybe, if I got it to full throttle, the problem would somehow go away. I put it in gear and started to drive the car while I attempted to open to full throttle in a gradual but fast way. As I got moving faster the windshield got soaked with raw fuel and the engine stalled. I figured out that once I got going fast enough, the fuel coming out of the carbs blew away and that starved the engine.

I needed help as I was clearly over my head. I went back to the shop and started calling around for help. I called Holley and spoke to an engineer who told me about the phenomenon of reversion. He had worked with Jim McFarland from Edelbrock before working at Holley and had a lot experience and understanding of what I was going through. Boy, was I happy to hear what he was saying, because I thought I might be considered nuts.

It would be years later that I fully understood the physics of what was really happening. In an intake manifold there are two significant events during the intake stroke of the engine, which are the opening and closing of the intake valve. When the intake valve closes, a compression wave forms that reflects the initial ram air energy up and down the intake runner much faster than the actual air flows when the valve is opened. These waves were the source of the energy that pushed the fuel out of the carbs before the next opening cycle. In a regular intake manifold, if the wave reaches a reflective surface, a wave of the same form is reflected back along the ports. Because the individual-runner fuel injector manifold didn't have any reflective area, and because I had mounted a carb at the entrance, these reflective waves were fast enough to push the fuel out of the carb before the next opening cycle took place and sucked the fuel back.

Now that I understood it more, what the hell was I going to do about it?

My approach was to build a bigger plenum area under each carb and make an H-pattern system of 1-1/4" hoses to connect all the individual plenums together. This would permit these compression waves, which mostly run along the surface of the

The changes to the Maverick's intake and carburetors required that a new hood scoop be fabricated in order to provide the proper clearance. This would give rise to the infamous and short-lived "eyeball scoop." Al thought the evil eye painted on the scoop was a nice touch. Others, including myself, disagreed.

port, to cancel each other out and to add to the reflective surface area of the bigger plenum.

I also had to have larger CFM capacity as the individual cylinders demanded more flow than what the 500-CFM carbs could supply. So I ordered two Holley Dominators, cut them in half, and mounted them. By making the plenum larger, each cylinder could then pull from both throttle plates. This increased flow while reflecting and dissipating the reversion. Bottom line was I now had a 9-second, 140-mph Maverick. I wasn't sure if I was more pleased with the increased performance or the fact I got it to work. At any rate I was now starting to get competitive with the big dogs.

The wider, longer opening of the new intake forced a change to a new hood scoop. I committed to getting another fiberglass hood and integrated the new configuration into it. I had to make it wider to clear the larger spacing from the two 4-bbl carbs to the four 2-bbl, but I forgot about the complicated throttle linkage that would be needed. After I made it, the linkage hit the sides at full throttle, so I needed additional clearance for that. I made a small bubble on the side that immediately looked like an eyeball to me. I got Ralph Hart to come in and do his magic. Abstract art was in at the time so we had eyeballs, lightning bolts, volcanoes, and such on the new scoop, which I thought was cool. However, it went over with like a fart in church with some, especially with my then-girlfriend Marge, who just hated it.

Coming home late from a race one night, we had to open the garage door as high is it could go to clear the spare tires that were mounted above the truck cab. Well, the garage door was not quite up all the way. As I entered into the garage the tires hit the garage door and pushed it up. It went up harder than normal, hit the limit of its travel and bounced back down just as the hood scoop passed underneath it. The door crushed the new scoop. Marge started cheering but I stood back and felt like a black cloud was still following me.

Here is Al in action with the Maverick at the 1970 NHRA Nationals. By this time the rules and technology in the Pro/Stock class were changing rapidly and the SOHC 427 engines were becoming long in the tooth and short in supply. The last major NHRA event won by a SOHC 427 powered Pro/Stock car would be the 1971 Summernationals.

I had to re-install the other hood with the ugly aluminum scoop on it.

By now I felt the car was running good enough to go to Indy for the first Pro Stock national event. It had been five years since I been there. No longer was it like the good old days of semi-amateur and/or individual team cars. Now it was nearly all professional two-car teams funded by sponsors. As I remember there were about 35 cars there trying to qualify for the 16-car field. It was painful that I was no longer going to be at the top of the qualifying ladder. I would have to learn to be happy to get into the mid-bottom of the 16-car field. Though I closed the gap I had more work to do, as I lost in the second round.

After the Indy race things in the racing world on the East Coast quieted down as the weather started to get cold. I pondered what was in store for next year. As I looked back on the year I realized that the disastrous clutch failure early on definitely delayed getting my new car sorted out and hampered the intake development. Most teams were getting dynos, performing ongoing engine development, and generally getting more professional.

1971: THE SUN SETS ON THE CAMMER

I started off 1971 with a really big bang, but this time a good kind of bang, by marrying my wonderful wife Marge on January 16 (as of this writing Marge and I have been together for 47 years).

Note that we got married in the off season, so as to not interfere with the drag racing schedule. For once my car would be the same as last year. I was able focus on R&D and engine development.

I only had two sets of aluminum cylinder heads left for the Cammer. I had 2 sets of steel heads that I really couldn't use because there were simply too heavy. They added about 80 pounds over the front wheels. I could still buy blocks and other components, but not aluminum heads. I realized that I had to start looking at alternative engines for the future.

Over the winter I took the engine apart and decided to open up the ports and polish them. I had never done that; before I just removed casting irregularities and matched the intake ports and valve radii.

GOING AROUND IN CIRCLES

Over the winter, I met an old friend named Walt Kaiser, whom I hadn't seen since the Bob's Big Boy days. Walt knew about me from all the Philly radio ads for the various drag races I was featured in. He said he was getting into dirt cars and wanted to know if I had any engines. I had successfully bartered an engine with Paul Deasey, also a dirt track racer, for work on the second-generation *Batcar*, so I said yes, I did. I would build him a steel-head Cammer, but this time it was for money to pay the bills.

Walt asked me if I would help him get the car/engine set up. I said sure, it sounds like fun. He even put the Mr. Speed "S" logo on the side of his car instead of a number (I found out later they called it the Shit Car) and my speed shop decal on the hood.

Walt was new to dirt racing and he hired a driver that he didn't know either. I was new to dirt racing too, and I wasn't sure what to expect.

The driver said he would take it slow around the track for a while to get a feel of the car. I'm not exaggerating when I say he was out there for more than a half hour, slowly idling around the track while the other cars were zooming by. I couldn't believe how slow he went and wondered how he could hope to learn anything going so slow. After he talked to Walt about the car I let him know about how I felt. He did go out a little faster the next time, but was still embarrassingly slow. I felt that he was scared of the car. Walt had for some reason put an automatic transmission in the car. When his driver did actually go slightly faster the second time out, the oil was forced to the side of the transmission oil pan. That uncovered the pick-up and burned out the transmission, and that was the end of the first day.

Walt put a manual transmission in it after that. We were back two weeks later and the "driver" was back to doing no more than idling around the track. When he came in I said, "Please, let me see you go sideways like dirt cars are supposed to." It was now apparent to me he was scared as hell of the car. I said any asshole could do what he was doing, and no one was learning anything, including him. He said to me, "Well, if you think it's so easy, why don't you do it?" I said, "No problem."

Because I didn't have my helmet and fire suit with me it was decided that I would be driving the car the next time out. You don't let your face

The only known photos of the Superman modified stock car. This effort was powered by one of Al's SOHC 427 engines. (Photo of car on track courtesy of Michael Shaub / Reading Fairgrounds historical society)

Certainly one of the best things to happen in Al Joniec's life in 1971 was his marriage to his lovely wife Marge. Our hero wore one of his Mr. Speed T-shirts to the ceremony to assure her she was not marrying any mere mortal.

write checks your ass can't cash. I figured I was used to going fast already, and I had been participating in dirt bike cross country racing, including a long Enduro race. Slipping and sliding while going sideways was not a big deal to me. I figured this would be a cake walk. WRONG!!!

When I started driving the dirt car it felt different but not bad. Blending into the traffic took some getting used to. I stayed low on the track as the faster cars were going high. I got used to the feel of the car, progressively going faster and faster. It didn't take long before I spun out and hit the wall. No matter what I did to correct the car, it was just tail happy and spun around. I went back in the pits and they changed some suspension settings. I went back out only to repeat the same kissing-the-wall theme. Bottom line, I didn't know what to tell them to do to make it better, and they didn't know how to make it better. I finally bent the back of the car and we went home.

As a side note, Marge and I were watching the race together when a big multi-car accident happened on the other side of the track. We couldn't see what caused it but one of the cars flew way up it the air, as high as the telephone poles, tumbling end over end. It came crashing down right on the concrete wall. It didn't bounce, but immediately stuck there. I said, "That's bad." The car landed upside down on the roof. The roll cage was crushed and the driver was killed instantly. They red flagged the race (which they hardly ever do) but never announced what happened. We found out later from word of mouth. This only emphasized the "What the hell am I doing here" feeling, for both of us. Nobody in our group knew what to do to set the car up properly for dirt. We did this one more week with the same results. I told Walt that this was nuts. He needed help and/or advice. He did get some help, bought some new tires to boot, and felt confident things would be better. And he talked me in trying one more time.

This time we went to a faster track in New Jersey. The car felt a little better to me during practice, but we had a long way to go. I was in the race field for the first time and in the middle of the pack when the race started. We went into the first corner side by side, engines screaming, dirt flying through the air, and cars bumping me from every side. I thought I could hardly keep up with the pack, let alone beat them. Finally a "holy shit" moment happened—my first really big accident, in a huge, multi-car pileup. I had one car directly on top of my roof dripping fluids on me. The rear suspension was broken and the brand new rear tires were punctured by another car ramming me from the back. Tow trucks picked us up and took us back to the pits.

I assumed that we would be going home after that, but no. Walt said that he could fix it and we would be back out on the track in no time. He simply repositioned the rear suspension struts and re-welded them in their old place and put the old tires back on the car. No weight checking, no alignment check. They just slapped it together

The 1971 NHRA Nationals would be the last hurrah for Al and Cammer power. By this point technology had outpaced the aging design that had served him so well since 1965. He also learned that he would be losing his only remaining sponsor, Rice and Holman Ford.

sent it back out. I was not so sure what to expect, but it actually felt like a new car. I couldn't believe it. The car felt like it was on rails and I was pumped. I was passing cars like they were standing still. I finally could drive it into the corners as fast as I wanted, and it turned without hitting the wall. I was flying on the high side of the track. It was actually fun and thrilling at the same time as I easily passed the fastest cars. I had to have been going more than 100 mph sideways in the corners as I was getting more and more comfortable

Then, after several fast laps. most likely going well over 140 mph, I hit a big dip at the end of the straight. It bottomed out the entire car and it felt like all hell broke loose. The driveshaft rear yoke had a heavy shield over it and when it bottomed out the yoke almost sawed right through the shield—and my balls—as I was sitting directly on this very shield. There was a huge noise and a big-time vibration that went straight into my package.

I immediately slowed down and went straight into the pits. I was greeted with all kinds of happy smiles. Everyone—but me— was thrilled that the car was going so fast. It was painfully apparent that Walt had no idea how to set up the car. I didn't know either and didn't care to spend the time to learn how, so I told him I wouldn't be driving it again.

That was the end of my dirt car driving career. As I look back on it now I remember the wild thrill of going sideways in the turns, steering with the throttle all the while throwing a rooster tail of dirt. At the same time I remember having to drive with one hand while the other hand wiped my face shield to clean off the dirt being thrown at me by all the other cars doing the same thing.

BACK TO DRAG RACING

I put the now-ported cylinder heads engine back in the car and I was, at best, disappointed. I was still stuck in the high 9s and needed at least 2 more tenths to be competitive with the Chrysler cars. Gaining two tenths of a second at this level of play was a big task. I wrongly assumed it was all in the horsepower department when I got a wakeup call.

One of the Southern track promoters asked me to run in an 8-car field of Pro Stockers at three separate race tracks that were in close proximity to one another. They needed a Ford in the field and they were offering guaranteed tow money. It sounded good to me, so I signed up.

The first race I lost to the eventual race winner, Don Carlton in the *Motown Missile*, predictably by about around 2 tenths. Dyno Don Nicholson was there too and we pitted together. He asked me if I wanted to try a 55-pound flywheel he had been carrying around for a long time. I needed to change something and thought it was as good as anything to try. The results were totally different than what I expected. I could now lower my rpms off the line—the heaver flywheel didn't let the engine stall. The big surprise was that when I shifted into the next gear it was like I got slapped in the back. The car would almost do a wheelie when the huge inertia of the heavy flywheel kicked in. I picked up around a half car length or so as it lurched forward in every gear. The second race was at a real slippery, dirty track where you had to back up to the starting line at a base of a mountain. I believe it was in Wilmington, North Carolina. The starting lights (note: not the standard Christmas tree) were hanging over and near the starting line, so it was hard to see them. The track owner said the solution was to look at the timing tower windows. The lights reflected off of the windows and it was easier to see from the line.

The reason why there was no traction was that the cars tracked dirt up to the line. Except for the actual track, everything else was unpaved. With only eight cars in the field there was no time to use rosin for everyone.

With the heavier flywheel, I had the advantage of using lower rpm to leave the line without stalling, and I had good throttle control. During the race the *Motown Missile* was paired off with me again. During time runs it was really apparent he couldn't bomb off the line (I believe the Chrysler guys used 3.1 first gears) and had to shift into second gear by the time they passed the starting lights. However, because I could now use a lower rpm off the line on the slippery track, I beat him off the line and he couldn't gain on me on the other end.

What a wakeup call that was. I won the race but needed a slippery track to do it. The secret was traction and not necessarily power. The next day we raced at a different track with good traction, and once again the Chrysler cars were the fastest. Even though I ran closer, I not only lost but blew my engine. I broke an aluminum rod yet again, which also took out another set of heads as well.

On the ride home all I could think about was traction and changing to a Boss 429 engine. The core problem proved to be that I was stuck with 12" tires while the Chrysler cars were using 14". As for the Cammer engine, I was down to my last set of aluminum heads. I knew I would be out of parts sooner or later, so change was the only option. In any case, I thought I had reached the limits of the Cammer years ago. Its 1.3:1 rocker arm ratio limited how high you could lift the valve open, which was less than .650". The Chrysler and Chevy guys were in the .750"+ lift range by now.

I found a set of real magnesium 16-inch rims from a dragster. They had 12-inch tires mounted on them and they worked much better. They weren't any wider, but because they were taller they had a longer contact patch (though I had to put a lower ratio rear gear in it to compensate for the larger diameter tires). I put together my last aluminum-head Cammer and went to Indy again, hoping for good results from because of better traction from the 16" tires. Once again there was a large field of 40-plus cars. And, once again, it was painful to acknowledge that all I could do was qualify in the 16-car field; it was not realistic for me to win. The Sox & Martin Chrysler team won again, and I found myself still about two tenths of a second away from being competitive.

I was already planning for next year on the trip home from Indy. I knew that wheel standing was not working going forward. You had to kind of drill off the line like you were glued to the track (like today's Pro Stock) without wheel standing. Unlike the Chrysler cars that were using a 3.1:1 first gear, I was using a 2.5 ratio rear. I had to spin the tires because the engine would bog if I didn't. I needed a 2.78 first with 14" tires to take it to the next level.

I finally decided that I would switch to a Boss 429, move the springs in and tub the wheelhouses for 14" tires, and lower the entire car starting immediately when I got home. My final surprise would end my season shortly after I got home, however. I lost my only sponsor, Rice & Holman Ford, and would now be a real privateer for the following year.

1972: AL GOES RACING ON HIS OWN DIME

The surprising news for the 1972 Pro Stock class was that NHRA started to handicap the field to stop the dominance of the Mopars. It was obvious to me that Bill Jenkins (with the help of GM) was instrumental in influencing these changes. Pro Stock had worked on a 7 pound per cubic inch rule until then. The new rules permitted 6.75 pounds per cubic inch for straight-valve wedge engines (i.e. small block Chevys) but the kicker was that the engine size was limited to 366 cid. This handicapped anything with hemi heads and canted valves and forced them to remain at 7 lbs./cid.

(Author's note: Dyno Don Nicholson related this story to me as well, indicating that Jenkins influenced NHRA because he was being beaten consistently by Ronnie Sox with his Hemi Mopar and Dyno with his Cammer Ford).

Jenkins immediately jumped on this and came up with a 331 cid engine combination that permitted his car to be as light as 2,234 lbs. Under the old rules the 429 had to be 3,000 lbs. This gave Jenkins a 750-lb advantage at least. Even if the smaller engine was way down in power, the lighter mass of the car more than made up for it. Jenkins capitalized on the rule change by dominating the class in 1972. I looked at doing a 352 cid wedge engine combination because I had such good luck and was so impressed with the Louie Tomaselli C/Gas engine. But because there were no tunnel ram intake manifolds available at the time (I incorrectly assumed you had to have one to be competitive back then), I didn't pursue it. I was again playing catch up with no help from Ford, and without a sponsor of any kind. And I didn't know I was already obsolete.

At the end of 1971 I took the engine out of the Maverick and delivered the car to my old and good friends Al & Red Rowe, brothers I knew from the Bob's Big Boy days. They had a welding fabrication shop and I trusted them to do a good job. They had done my truck ramps and storage in the past and I loved working with them. I told them to move the rear springs inboard and tub the wheel wells to fit 14" tires inside the fenders. I also wanted to lower the whole car as much as practical. Unlike some times in the past, I told them to take their time and call me when it was done.

I could then focus on the new Boss 429 engine combinations. By now most of the best Ford Pro Stock cars were already going to the Boss. There was a tunnel ram available for them and that was all I needed to make the leap. I didn't want to go through the intake manifold development thing again if I could help it. I had to do my homework on every part and order everything from scratch to make sure I was getting the latest and best parts available. A lot of R&D was going on with the Boss 429. I was really pleased to see that JR headers were making a set for a Boss in a Maverick.

I couldn't believe how heavy and robust the block, heads, and crankshaft were compared to a Cammer. Speaking of heads, the ports looked

Al found himself racing as a privateer for the first time in years as the 1972 season dawned. He debuted the Maverick, now painted in basic black and powered by the Boss 429 engine, at Atco Dragway.

like sewer pipes, they were so large. It really was a totally different engine and I had a lot to learn. Again I went with aluminum rods as they were readily available and the steel ones looked way too heavy. I had the very large Boss crank rod journals turned down to the Chrysler size, which helped lighten the crank as well. Pistons were available but you had to have the valve clearance cut according to the cam specs. The rockers were unique to me, with the very long exhaust and short intake rockers, but at least they were now 1.75 ratio. With the tunnel ram manifold I used two smaller 600 CFM carbs, at Wayne Gapp's recommendation. This was a combination he was already using successfully. I had helped Wayne with his Cammer when he had a small problem and we had established a working relationship.

I got the car back, all tubbed and lowered, put the engine in, and took it to the track to test. The car went straight and true with no vibrations, and with the 14" tires it really was much better off the line. I was disappointed with the results, however. I was barely getting in the high 9s at only 137 mph. The engine felt lazy as it simply didn't accelerate like the Cammer in 1st & 2nd gear but it felt OK in 4th. So here I was with a much better car but felt like I was starting all over again with the heavy rotating mass of the Boss.

I knew that the street version of the Boss 429 was no match for a standard 428 Cobra Jet, and I was starting to understand why. The Cammers were an evolution of the FE engines that started out as 352/332 cubic inch engines that ran very well. At 427 cubic inches the cylinder walls became too thin, and higher rpms took a toll on the oiling system (requiring a change to side oiler), and the rotating mass of the piston rod assembly were now becoming marginal. The fuel Funny Cars were going through Cammer engines like firewood.

The Boss 429 was created to correct these problems. They were robust by design but the downside was a very heavy rotating assembly and heavy block and heads. Its huge ports were the result of the "bigger the ports, the better," mentality at the time. The bore to stroke ratio was much better for the 429. The bigger bore also unshrouded the valves and could be enlarged to well over 500 cubic inches easily. IHRA Pro Stock teams were doing just that, running so-called mountain motors (unlimited cubic inch engines) in match races at that time. The Boss engines were very popular at 600-plus cubic inches of displacement, as this made the huge ports more closely match the engine displacement. I found myself stuck with yet another development task, but this time with no sponsors. I was pretty sure I had a good car, so now I was totally focused on the engine. Because I didn't have a dyno, I could only test at the track while racing.

The engine combination that I started with was only good for around 7,500 rpm, so I knew the ports

Joniec Over Rowley In Cecil County Pro Match

CECIL COUNTY DRAGWAY, Md. — Cecil County Dragway, one of the most popular racing complexes on the East Coast, held their 1972 Grand Opener this Sunday with fans and racers alike making the scene after the long cold winter. The featured event of the day was a special match race between two of the finest Pro Stock drivers in this area, Al Joniec and Vern Rowley.

Joniec was at the wheel of his new Boss 429 Maverick, which he had just completed in the last three weeks. Rowley was at the helm of a very familiar car, that being the hard running Plymouth Duster owned by Bill Stiles. At the '71 opener at Cecil County, Rowley hung three straight losses on the Shallcross Brothers' Camaro, and he was out to do the same at the "Super '72" opener.

Round one opened up with both drivers making a pair of super burn-outs, which had the spectators on their feet in anticipation of a great match. Al Joniec jumped into the favorite's role immediately when he ripped off his best time ever at 9.72, 140 mph. Rowley ran a sub-par 10.20, and he knew he had his work cut out for him in round two.

But when round two rolled around, it was almost the same story as the initial round of competition. Joniec's Boss 429-powered Maverick put down another great run, this time stopping the clocks at 9.82, 143 mph. Rowley once again had some problems, as the usually potent Stiles' Duster could only hit another 10.20 elapsed time. One thing is for sure, Vern Rowley will be out to even the score when Cecil County Dragway hosts an eight car Pro Stock Spring

Cecil County Dragway announced a match race featuring Al Joniec's Maverick vs Vernon Rowley's Hemi MOPAR as part of their 1972 Grand Opener in March of the year. Al prevailed over Rowley, posting a best time of 9.72 seconds at 140 mph.

June 2, 1972 — National *DRAGSTER* — Page Twenty-Three

Joniec's 9.70's Good For Atco Pro Victory

ATCO DRAGWAY, Atco. N.J. — Al Joniec's storming Maverick, fast becoming one of the nation's swiftest Pro Stock Fords, was the big winner in Atco's four-way Pro Stock match race, besting the likes of Bobby Yowell, Rich LaMont and Sam Auxier with strong 9.70 clockings.

In the first round, Auxier's Mustang scored an upset win over Bobby Yowell's Duster, 10.07, 137.82 to an 11.09, 90.45. Yowell blew his motor on the run, no doubt contributing to his sub-par performance.

Joniec then came up to the line and scored his first win of the affair as he ripped off a great 9.73, 142.40 to handily defeat LaMont's Gremlin at 10.09, 137.61.

In the second round, it was Auxier's turn to dump LaMont, cranking out a 10.07, 136.98 to the ailing American Motors machine's 11.31, 87.46. Joniec soloed with a great 9.72, 141.73, which turned out to be low elapsed time of the meet.

The third and final round saw the two undefeated Fords of Joniec and Auxier go at it for the overall title. Joniec won decisively with a 9.80, 141.95 mph thrashing of Auxier's 10.05, 138.03. LaMont made a solo with an 11.51, 84.98 mph clocking to end the action.

In other action, Jim Hill's C/Dragster turned a 10.31 at 132.74 mph to defeat John Perotti's B/Dragster at a red-lighting 9.01, 133.23 for the Top Eliminator title. Doug Wiley was doing the driving for Perotti, and it was only his second time out.

Street Eliminator was captured by Randy Mason's C/Gas '70 Camaro with a 10.35, 132.54 mph conquest of Herb Richman's D/Gas Chevy II, losing with an 11.17, 123.11.

In Super Stock Eliminator, Jack Worrell's SS/KA 1968 Ford, which had just set a new National Record for MPH at Englishtown, turned in another stellar performance to win the title with an 11.83. Runner-up honors went to the Franks & Orsi SS/LA Camaro at 11.92, 104.89, breaking out on the final round.

The popular Stock Eliminator bracket was taken by Paul Stein's B/S Camaro with a 13.58, 102.38 and Factory Experimental Stock fell to the Boorman & Lillie Falcon wag at 11.88, 116.73 over John Hoffman's "Nitro Nova" at 12.25, 112.07.

ATCO DRAGWAY, Atco, N.J. — Al Joniec's Ford Maverick was unbeatable in Pro Stock action, hitting a best of 9.72 at 142 mph to defeat Bobby Yowell's Duster, Sam Auxier's Mustang and Rich LaMont's Gremlin in a 4-way match race.
Photo by John Kinker

In early June 1972, Al Joniec and his Maverick proved unbeatable in a Pro/Stock event at Atco Dragway. Al dropped the Hemi Duster of Bobby Yowell, the AMC Gremlin of Rich LaMont, and Sam Auxier Jr's. Maverick in the final while posting low E.T. for the event with a 9.72 at 141.73 mph.

were way too big. I also knew intuitively that the two 600 CFM carbs were inadequate. I upgraded to a bigger cam with heavier valve springs, put two Dominator carbs on, and started to reduce the port size by epoxying the floors of the intake ports. This brought the power up in the 8,000-plus rpm range and it picked up a lot. I remember Wayne Gapp looking at my Dominator carbs and shaking his head but he also changed his carbs shortly after that. I was now starting to go in the 9.7's at 142 mph. That was around where my best Cammer times were, but I was still 2-plus tenths off where I wanted to be, as I knew I had a better car with 14" tires. However because of the Pro Stock rule change I was off more like 3-plus tenths if I wanted to win any big races and/or beat Jenkins. I was winning some local races by June, though.

Al improved the look of his Maverick by adding flames to the paint early in the year but the big improvements came from the improved cylinder head ports and aerodynamic improvements such as enclosing most the grille with Plexiglas, removing the spoiler from the rear deck and adding a chin spoiler to keep air from getting under the car. Al felt he had a chance in the Pro/Stock class at the 1972 NHRA Nationals. After qualifying well and going rounds in competition, it became apparent that the trend toward 6.75 lbs. per cubic inch of displacement weight-break cars would be the only way to win going forward. The Maverick and Boss 429 were doomed to extinction.

Al was able to close out the 1972 racing season successfully at local events and match races, where his combination still remained competitive. It was at Suffolk near the end of the season that he posted his best performance of the year with a 9.48 at 146 mph. The photo of Al's Maverick leaving the starting line at Atco late in the season shows the radically different stance to the car due to the modifications required to keep it competitive in the quickly evolving Pro/Stock class.

While talking with Wayne Gapp, who was doing the high-torque, mid-rpm thing and was successful with this combination at the time, I learned that he was reducing his port size to really make it work. He and Jack Roush were Ford factory engineers and were further along than I was. They were continuing to do ongoing R&D on the heads. They offered to sell me the latest and greatest set of heads they developed, so I bought them as I couldn't afford to waste any more time. I now felt ready to go to the Indy Nationals. The car had picked up more power with these heads and I started to think about aero drag. If you blow up a picture of the car you can see the clear Plexiglas I used to fully enclose (except for a 10" section) the grille. I continued to fill the grille area below the bumper, hid a thin, flexible chin spoiler near the front crossmember to help keep the air from going under the car, and sealed the hood with foam incrementally one weekend at a time. I had already removed the rear wing spoiler and continued to lower the front of the car a little bit at a time to get the best aero stance I could. After all of this I saw a 2- to 3-mph improvement with the same power. I was now able to compete with and beat most of the other 7lb.-per-cubic-inch cars, but the lightweight 6.75lb cars were still faster. However there were still far more 7-lb cars than 6.75 and match races were still good places for me to race. I did a lot match racing after the Nationals. I went down south again and was kicking ass and taking names. While winning one last match race I went 9.48 at 146 mph (if I could have only started out that way). It was the very best time I turned, but I broke a rod. Unknown to me at the time, this would be the last time I raced the car. The bottom line was that the car was obsolete as it had now become the era of the Pinto in Pro Stock.

I was pondering what to do next because I was running out of money and had no sponsors.

Right around this time a guy I knew of but never formally met, named Sam Feinstein, walked into my shop. Little did I dream what would happen next.

CHAPTER NINE

ENTER SAM FEINSTEIN AND OLLIE THE DRAGON

I DIDN'T REALIZE at the end of 1972 that I was getting burnt out from the constant 24-7 pace, with no weekends off or vacations, that had been my life up until that time. Most of all I was trying to figure out how I could continue racing without a sponsor. I simply didn't have the money to upgrade to a new car.

Then Sam Feinstein walked into my shop.

Sam was a sports car driver/owner from the Philly area who raced in the Sports Car Club of America (SCCA) with a real 427 Ford Cobra. He knew about me and my Ford history; I had heard of him but we had never met. After he introduced himself, he said he would like to work with me somehow. "All I want from you is to take care of my Cobra," he said. I was taken aback but also pleasantly surprised, so I asked him to tell me about his car. Before he could answer his well-endowed trophy girlfriend spoke up and said the car was called Ollie. I said, "That's a neat name, but why do you call it that?" She said it was because every time they first started it up, fire came out of the hood scoop. So they dubbed it Ollie the fire-breathing dragon.

I said, "OK, I think I can help you." I got the feeling that Sam was a genuine guy and all he wanted was to race with a competitive car. Plus, he had the resources and was willing to properly fund the project. I said it sounded doable but my fee would be a Pro Stock Pinto and truck for me to race in exchange. He would own the operation and could sell everything in the end to recoup some of his investment. One more stipulation I had was that he had to buy a dynamometer so I could do proper development work for both the cars. It was a win-win deal and our contract was a handshake. All of a sudden here was a solution to my problems. For the first time in a long while I was excited and looking forward to the new year. We agreed to have me move into Sam's facility; I became the general manager of Feinstein Racing. I wanted to do a good job for him so I jumped right in with both feet. I didn't have anything but an idea in my head regarding my car, so I started on Ollie first. I put it on the lift to understand what I had to work with. I was appalled at the very small ball joints (Cobras were really English-made AC Bristol 4-cylinder cars with Ford engines shoved into them). I'm sure these were good enough for a lightweight car with proportionally small tires for street use, but they were not adequate for big-tire, pro-race applications.

The next point of concern was the solid brake rotors. They had no vent holes of any kind to cool the brakes. The energy of a car going from, say, 150mph down to 50 mph, converts a huge amount of energy to heat. If you don't and/or can't get rid of the heat at the same rate you make it, you get brake fade and can even boil the brake fluid and the car can't stop. In road racing, brakes are like reverse horsepower; as with engine horsepower, you need all you can get.

The frame was nothing but two 4" tubes running from front to back with 4" cross members and everything hung off of it. The rear end was bolted to the center of the frame rails with U joint swing axles. I could see that the frame had been damaged and repaired many times. It was soft as hell as I could easily cut it with a knife. It did have independent rear suspension with A-arms, and both front and rear A-arms looked OK. Hell, they had to be OK, as all the parts needed to be stock according to the rules.

I then turned to the engine. The engine in the car turned out to be nothing but a stock engine. I asked them to show me what they were using for an ignition timing curve. They showed me an old book that looked like it came from Shelby. It specified only 6 degrees of initial timing. I then understood why there was fire coming out of the hood scoop when they started it cold.

After installing the dyno in the shop, the first engine we put on it was Ollie's. As I recall it felt very lazy. It was just under 400 hp, but now I had my baseline. Again I had to go back to the SCCA rule book to see what was legal. The rules specified stock blocks only (no aluminum), a max of .030" overbore, and stock type replacement pistons. That put us at 430 cubic inches. I believe there were several stock head numbers that were listed, but they also had to remain stock. Intake manifolds were free, so you could use anything as long as it had the stock 780 CFM 4-barrel carb, but without any carb-type adapter to raise it up. Cams and valve train were also unrestricted as was the exhaust. Road racing called for excellent mid-rpm torque to pull away from the many corners. Some hairpin turns needed to be taken all the way down in first gear. So lower-rpm grunt was required. There were also 2-hour endurance races, so reliability was a major consideration. I had to limit the engine to a lower maximum engine rpm than I was used to for drag racing.

I started to order parts to build a complete second engine. I bought as many different intake manifolds as I could, both new and used. Eventually I wound up with six different intakes. I also got all three sets of heads that were listed as legal. Now that I had a dyno I didn't have to guess at

Sam's Cobra, affectionately known as Ollie the Dragon, is seen here in its early racing configuration looking much like any other 427 Cobra of the era. Al Joniec's first task with Feinstein Racing Enterprises was to get Ollie's 427 making the power needed to win.

Along with getting Ollie's horsepower output up to speed, Al was instrumental in modifications to the car's chassis and body work.

what worked best. I ordered two different cams to start with and would go through at least 3 more before I settled on one. My initial goal was to keep the engine under 7,400 rpm with a minimum of 600 hp and 500 lb-ft of torque at 3,500 rpm.

Lightweight wrist pins were just coming out. I jumped on them as I wanted to reduce the reciprocating mass as much as possible. I ground off the casting flashings of the sides of the rods and polished them to help the cause, while drilling some lightening holes inside the pistons myself, weighing them, and making them all the same weight as I went.

I had the block honed and decked and the crank assembly balanced. I did many dry trial assemblies to check for valve-to-piston clearance, deck height, and also put the head on to check for valve shrouding. On the heads I did classic thin valve seats. I was sure to keep the seats at the very edge of the valve, with a minimum of 2 gradual angles approaching the valve from within the valve pocket and one angle in the combustion chamber. Both sets of heads were cut to the minimum CC's , including the intake manifold side. I made up my patented end supports for the rocker shafts, and threw away the shaft springs and replaced them with aluminum spacers.

Finally I put it on the dyno and after a little break-in time I made a pull. The engine made around 560 hp at only 6,800 rpm. I thought I was too conservative in the cam timing so I installed in the next bigger cam. The power and rpm went up to 590 hp at 7,200 rpm, so I knew I was going in the right direction. I ordered two additional cams with more lift and duration. While I was waiting for the cams I started to test manifolds. I had three 360-degree open types—Weiand, Offenhauser, Edelbrock—and three 180-degree types—standard Ford, Ford Sidewinder, and Edelbrock. I started with the standard Ford manifold and standard 427 heads (with smaller ports). I tried the 360-degree types first and they were either down on power or lost low-rpm torque compared to the standard 180-degree Ford. I felt good as I was learning what worked without the trouble of going to the track. Even if we had tested at the track I'm not sure I could have told the difference with lap times over one minute.

The bigger cams came and I went back to the baseline intake to test again. This time the power went up to 610 hp at 7,400 rpm but I was looking for more. With the next bigger cam the power went up again to 625 hp, but the lower-rpm torque went down. I talked it over with Chase Knight at

Sam Feinstein takes a victory lap after winning yet another SCCA race with his Cobra. With the assistance of Al, Ollie the Dragon had become a winner.

Crane cams and we agreed to use the same cam lobes but open up the lobe center lines to 114 from 108. While I was waiting for the new cam I tried the 180 Edelbrock intake and it lost a little power in all the rpm ranges. This testing took place over a period of months while we continued to race the car. The engine came in and out routinely. While checking the suspension settings I noticed that when Sam came back in from the track trials the settings were not the same as they were when the car left the pits. I thought the frame was bending or flexing. Sam appeared happy; he was winning races with the new engine and he claimed he had to learn to drive all over again due to the newly found power. But this was only the beginning. I thought we should strengthen the fragile Cobra frame by adding a 1 3/4" chrome moly subframe bridge from the rear roll bar area of the frame, through to the tops of the doors, inside and across the firewall, and down to the front suspension area of the frame. This would help keep the car from twisting and would help protect Sam by adding a roll bar on the side.

I didn't like the dated, drab, dark green paint job the car wore. I talked Sam into letting me update the car with a new paint scheme. I made up a full-face spoiler and mounted it. I also laid back a full-width windshield (in green-tinted acrylic) to reduce the aero drag. I took out the headlights and used the holes to provide ram air for the engine via hoses leading to an air box. I had learned in 1963, when I first put on a fresh air ram kit, that I got unbelievable gains of 2 tenths and close to 4 mph while doing nothing else. After that I realized I couldn't afford to be without ram air ever since. The factory hood scoop on the Cobra was non-functional as it was too close to the hood. It lay within the normal surface-air boundary layer, so at speed air actually went over the scoop, not into it.

I continued to work on the engine development while the car raced in the 10 or so SCCA regional races in the eastern division. You had to win your division or be runner up to be eligible to run for the championship at the end of the year. We were taking care of business; I don't believe he lost any races, except for one time when we had a drivetrain failure. Between races I would update the engine with the latest things we learned on the dyno. The latest cam with wider 114-degree lobe centers worked fantastic. We now not only had big power, but also great torque at 3,500 rpm. After all these years I finally tested my patented rocker arm shaft end support brackets on a dyno. They were good for 20-plus horsepower, but their effectiveness was heavily influenced by the cam lift and valve springs. Speaking of rocker arms, I got a set of Iskys and they proved to be better by around 12 hp when compared to the OE rockers.

Following the successful 1973 SCCA season, Ollie the Dragon was commemorated in this 1/24th scale model kit produced by Revell.

I checked the actual lift with both stock and Isky rockers at the valve and it was around .020" higher with the Iskys. Apparently all the Iskys were closer to 1.75 ratio than the stock ones, which were inconsistent—some were as low as 1.7.

When I did my dry assembly of the engine with only the heads on to check for valve shrouding, I thought that I would one day relieve the block. The stock head gaskets already had a relief in them for this purpose. I used the gaskets to scribe the block, to be sure that the the sealing area was unaffected. I then ground a fairly large scallop-shaped path for the intake air to flow around the valve area nearest the block, no lower than the top of the ring land at TDC. This modification added almost 15 horsepower. I tried a Cloyes double-roller timing chain to replace the stock piece and was careful to make the timing setting the same. I was pleasantly surprised that it picked up another 6 horsepower. I originally made the switch to make it easier to change the cam timing, but I had to assume the rollers were much more fiction-free and it showed.

The best intake by far was the Ford sidewinder. I guess the Ford engineers knew more about these things than most gave them credit for. I found it to deliver what a true road-race intake was supposed to, good peak power and a wide powerband. Out of the two legal part-number heads that I could use, the smaller-port, standard 427 heads were much better than the larger-port heads. The standard heads matched the sidewinder intake port size as well. Again the Ford racing engineers knew what they were doing.

By the end of the year the engine was now in 660-plus hp range and was reliable.

One particular occasion at Pocono Raceway was a proud day for me. Pocono is a 2-1/2 mile track with a long straightway that allowed you to safely get your car close to top speed. The road race course wound through the infield and then onto the very long straight with its heavily banked turn. That day the heavily funded, factory-supported Corvette of John Greenwood was there. He had just entered the long straightaway section of the track as Sam entered a few bus lengths behind. As they both entered the straightaway I watched intensely as Sam slowly started to close in on him. Sam passed him before the braking point near the end and I was as proud as a father with a newborn child. I then felt we actually had a shot at winning the championship.

When Sam came into the pits he complained that the car was misfiring in a way that he never experienced before. After quizzing him about things, I asked him if he looked at the tach. He said he had, indicating that it read somewhere around 7,400 rpm. I did some calculations and figured he was going more than 200 mph. At that point it was hitting the rev limiter (got to love those type of problems). We had the fastest car that day but we lost the race when the ring and pinon gears broke. They had worn prematurely because of the extreme high speed and high heat of this type of fast-track racing. We should have had a differential oil cooler, but we hadn't needed one at any previous racetrack.

Sam went on to win the rest of the season races and won his division for the first time. As a result we were eligible to go to the SCCA Championship in Road Atlanta that year.

By 1973 the Cobras were no longer competitive against the Corvettes. The Shelby team had not been in competition since 1965. Therefore, no development work was being done on the Cobra; Ford had placed its emphasis on the GT-40. As a result there was a sea of 454 Stingray Corvettes everywhere we went. GM was doing all kinds of R&D with the various teams out there, such as John Greenwood. I believe there were around 20 Corvettes at Road Atlanta and only 2 Cobras. Besides all of the modern suspension updates, the Corvettes had larger 454 cubic inch engines that were being heavily developed in all other forms of racing as well. The Stingrays were very aerodynamic and they didn't have to use solid brake rotors and other old technology as we did with the Cobra.

The Road Atlanta championship drew all the top teams from across the country that we had never faced before. Suddenly we were confronted with professional teams with big rigs, some with multiple cars for different classes. It was big-time, and it was intimidating. Between all these cars fighting to qualify for the starting field and lots of time trial runs available for chassis setup and to allow the drivers to get familiar with the track, the pace was hectic.

Because of the large number of entries, just getting new tires for the race was a task. You needed one set for practice and another set to race with. The tires designated for the race were supposed to be "scrubbed in," which meant taking them out for a few laps to make sure they were ok and to remove the casing marks. These sets were saved for the race, as you didn't want to go out with brand-new, sticker tires when the race began. This was the big time/ big boys, "wine and cheese" crowd. The whole thing was new and exciting. I loved the action as it was like a breath of fresh breath air to me compared to the drag racing that I knew so well.

We qualified on the pole and we were ready to go for the big race. There was a lot on the line for this race. Sam was a cool kind of guy who was hard to read when the pressure was on, but everything was going good so far and he seemed fine. I just was worried he might psych himself out, so tried to get his mind off the race by telling BS stories while we were waiting for the start.

Being on pole position, Sam was already in the lead when the race started. But it's hard to estimate just how hard to go to keep everyone at bay. In many ways it's easier to be in second place fighting for the lead than being out front, but he maintained his lead lap after lap. Sure enough near the end the race he was challenged by the second-place Corvette, who passed him strategically on the last lap. I couldn't believe it. I thought for sure that Sam would pass him back, but there wasn't enough time. I was in shock and the entire crew was disappointed. We were so close. Sam told me that by the end of the race his brakes were almost totally gone. At that point he was simply being out-braked by the Corvettes.

Totally depressed, I went over to the winner's circle, where the top cars were waiting for the post-race tech inspection. I went over to the winning Corvette to look at the engine. There for everyone to see was a stack of large spacers under the carburetor, which raised it up and created a small plenum area. I knew that it was illegal—I wanted to do the same thing as I thought there was a potential 10 to 20 horsepower in there, but I had followed the rule book. I grabbed the tech inspector, identified myself, pointed to the spacer, and informed him that was illegal. To my surprise he said yes, indeed it was. He told the Corvette guy the bad news and disqualified him on the spot. I expected different, but the Corvette guy didn't even argue and accepted the ruling (I think he may have known he was illegal). Sam was then declared the winner and all of our moods changed. Life was good again and it was party time.

Ours was the first Cobra to win a major sanctioning body national Championship since the Shelby team campaigned them in the 1965 era, and the last one to win a title to date. At the time it truly was the fastest Cobra in world.

OLLIE THE DRAGON'S OTHER LIFE

Cobra authority Carl Mentz kindly filled in the blanks in the history of the Cobra, designated CSX3009 by Shelby American, that later became famous as Ollie the Dragon in our story. Essex Wire Company was a supplier to Ford Motor Company and in 1965 the president of the company decided to form a racing team. CSX3009 was purchased in April 1965 and drivers Skip Scott, Dr. Dick Thompson, and Ed Lowther took her to fourth place nationally that year. In 1966 Lowther took the car to a national championship in the A/Production class and followed that up with a second-place finish to another Cobra in 1967 before the car was sold to Sam Feinstein. The car still exists and has been restored to how it appeared in 1965 while wearing Essex Wire livery.

In a previous incarnation, Ollie the Dragon had been successfully campaigned as the Essex Wire Cobra. The now-restored car carries this livery. (Carl Mentz)

Coinciding with my association with Feinstein Racing and the Cobra, I was in the middle of choosing a builder for my new Pinto Pro Stock car. S&S was building Bill Jenkins's cars and they were close by, so I thought I would be doing business with them. However, Fred Faulkner of dragster fame somehow heard I was looking to have a Pro Stock car built from scratch. He wanted to diverge from the dragster building business and jump on the new Pro Stock opportunity. He put the hard sell on me and made me a deal I couldn't refuse. He too was close by, just outside of Philly in Wilmington, Delaware. Even though he had no experience in Pro Stock, I thought he might be able to carry over some innovative tricks he had learned while building dragsters.

I blame Bill Jenkins for starting the trend of building cars from scratch. At the start of 1972, 99 percent of the Pro Stock cars were equal

During his affiliation with Feinstein Racing Enterprises, Al contracted with Fred Faulkner, a well-known builder of dragsters, to construct a 1973 Pinto for Pro/Stock competition. The Pinto was based on a custom chassis using only the Pinto roof and quarter panel sheet metal. Al's Pinto Pro/Stock entry was painted in a livery similar to that of the Feinstein Racing Enterprises Cobra. Both were Al's designs.

The FRE Pro/Stock Pinto charges off the starting line at Atco before spectator-filled grandstands.

except for his. Unlike the old Maverick, which was simply a new car off the showroom floor that had been modified, the new Pinto would be a different story. I ordered just the sheet metal for the roof, rear quarters, and doors and bought the fiberglass front end. The rest would be custom-built from chrome moly tubing, and there was no requirement for a VIN number. The sheet metal would dictate the size and shape of the tubing (all handmade blended bends and individually tig welded) and the frame would hold the rest. Most don't understand the amount of money, time, and energy it takes to build something like this from the ground up. Just the responsibility of precision positioning of all the components in a structurally sound and safe way requires skillful craftsmanship and hard work. While Fred was building the car, I was working on the engine. Here I found myself with a 351 Cleveland motor— yet another new engine I had to learn about before I could even start to buy the parts. As far as I knew there was no so-called school of thought these engines, so I had to learn the hard-knocks way. I went with two

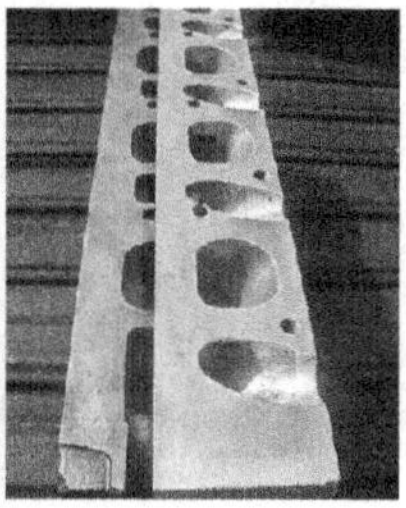

Al developed two engines for use in his Pro/Stock Pinto. One was based on a Boss 302 with a quarter-inch increase in stroke, which brought it to 331 cubic inches. The other was a 351C that was considered the norm for Ford Pro/Stock racers. Capitalizing on what he had learned with the Boss 429, Al decreased the size of the Boss 302 cylinder heads' cavernous intake ports and used a set of Jack Roush-developed plates to raise the exhaust ports. Al and Jack Roush remain friends to this day.

engine configurations, one a 351 Cleveland the other a 331, which was a regular Boss 302 with a 0.025" increase in stroke.

The main difference with the 351 Cleveland was the height and weight of the blocks. My belief was that you can typically get more horsepower per cube with a smaller engine and lighten up the car around 125 lbs at the same time. I put together a regular bottom end using standard TRW pistons with aluminum rods for both engines. However, the ports in the cylinder heads were too big again, as they were on the Boss 429. So I filled about 1/3 of the intake port floors (even more for the 331) with epoxy to change the angle of attack and increase the intake velocity. I would experiment with different port sizes over time, which was easier to do with epoxy. On the exhaust side I bought some aluminum plates from Gapp and Roush to raise the exhaust ports. These plates also changed the angle of attack and reduced the sharp bend for the exhaust flow.

CONNECTICUT DRAGWAY, East Haddam, Conn. — Former Winternationals champion Al Joniec displayed his best form in quite some time as he outlasted a strong Pro Stock field by defeating Dennis Ferrara's Camaro in the final with a 9.22, 148.27.
Photo by Clayton Taylor

Ivanof, Joniec Win Connecticut Opener

Al's 1973 season started off well with a Pro/Stock class win at the York U.S. 30 invitational, where he posted a 9.62 at 144.46 mph. He followed up with a win at the Connecticut Dragway opener with an even quicker and faster 9.22 at 148.27 mph.

I was pleasantly surprised to get around 620 hp with the 331 and 650 hp with the 351 right out of the box. However, when I put in the latest and greatest cam, which was supposedly what some of the other guys were using, the valve springs began to fail. You have to remember the material science for good, reliable springs didn't come about until around 1994. We were starting to lift the valve around .800" at around 8,700 rpm (with 9,000-plus over-speeds) and the valvetrain technology at that time was being pushed to its limits. I was now getting into the 675hp area but I was breaking 1 to 4 valve springs every test (what a pain in the ass). I was forced to try different valve springs in an attempt to correct it.

During this time I was racing the car as well. At the track after every run I had to take the valve covers off to do a spring inspection. Adjusting the valves is bad enough—how about changing several valve springs after each run? I had to carry many spare springs and if I ran out, that could determine whether I finished the race or not. It was frustrating, though it appeared to be the next phase of running a Pro Stock car as I wasn't the only guy having the problem.

One of my failures was the 331 cubic inch engine. The engine did have a good horsepower to cubic inch ratio, but I had to take the 125 lbs out of the car to make it effective. Unfortunately all my extra weight was in the rear, and when I took it out the car just didn't have traction. I took a regular steel rear bumper and filled it with lead to add as much weight (around 75 lbs) as I could to the very rear of the car. This was a trick I learned when I was working on Bill Jenkins's 1965 car when we filled the taillight housings with lead. I then painted the bumper white to make it appear like the front fiberglass one, and as similar as possible to the fiberglass rear bumper that I used with the 351 package. I still remember when one day some guys were helping me push the car to the line and one of them felt the lead in the rear bumper and gave me that OMG look.

Speaking of having someone to push the car, I never had a formal, full-time crew to help. I did have Dave Stillwell, who worked full-time for me at the speed shop, but he never wanted to go to the track for some reason. I did have some temporary

At the 1973 NHRA Nationals, Al qualified for the tough Pro/Stock field in spite of damage to his engine caused when Mallory metal dislodged from the crankshaft. While he had gotten progressively quicker throughout the year, a two-tenths of a second disadvantage to a racer named Bob Glidden at that event proved to be the impetus for Al, along with a number of other drivers, to abandon his efforts in the Pro/Stock class.

help at the track once in a while, but most of the help I got was from friends and sometimes even customers who wanted to experience what racing was all about. After 1969, the only person that I could really count on was my wonderful wife.

Near the end of the year it was just me and her. We both worked late into the night preparing for a race in upstate Pennsylvania, and we left early in the morning. She shared driving the truck as well, but we were both so tired that I had to pull into a rest stop and we both fell asleep in the sleeper part of the cab. We woke up to all kinds of people talking loudly. When I got up to investigate what all the commotion was, I was surprised to see around 50 guys in Army uniforms, as we were totally surrounded by a traveling convoy. All of these young guys had their noses on the window, looking in. I rolled down the window and asked them what was going on. The truck was a fully enclosed one with no markings of any kind to identify it as a race car truck. They were laughing their asses off, as they thought we might be going at it in the back and were there to cheer us on.

That was the last funny thing that happened that day. We woke up and drove to the track. Both of us were still very tired, and my wife was not feeling good. The Race Gods were not with us that day and we lost. I remember that, while driving home, I started to look objectively at my priorities for the first time. Either I was going to commit to continue and change something, or accept that my racing days were over.

I did go to the Indy Nationals once more. I was now getting over 710 hp on the dyno and starting to go into the very low 9s and I wanted to give it another try. I built a new bottom end for the attempt and put it in the car. On the second time trial run I had an engine explosion of sorts. What had happened was that the heavy metal that is used to balance the crankshaft (kind of like a 1" x 3" round metal plug) was centrifuged out of the crankshaft even though it was spot welded in. It went completely and cleanly through the side of the engine block, hit and badly bent the starter motor, and stayed stuck between the block and starter. I was surprised to find the plug, so I took the pan off and knocked the plug back into the original hole in the crank. As for the hole in the block, I found a wooden broom handle that was conveniently near the same size as the hole. I cut off a piece big enough to go through the block without sticking outside and I epoxied it in place.

The wood would expand and help seal the opening when wet. I then found a welding service at the track and had him come over and weld the heavy metal plug into the crankshaft properly. From that day on we called it the "broomstick motor," and it was with me to the end.

I did qualify in a heavy field of cars, but I was still about 2 tenths off the top contenders, including a guy named Bob Glidden. He not only won at Indy but he also won the Winternationals with Wayne Gapp's old car (Glidden beat Gapp's new car to add to the insult). As a result I felt I didn't have a decent chance of winning like I had in the past. I realized that most of the faster cars were now being heavily developed with a full-time, in-house engine dyno program that had full-time crews with outside help when needed. This forced me to look in the mirror and reassess. I realized that I was only BS'ing myself into believing that I could do two cars in two different series effectively in the same year. At least Sam's Cobra won his championship.

I continued to race and finished the season. I sold the car to my used race car lot, Carmel Ford in New York. This would be my last year in drag racing. After close to 15 years of racing, I was now 34 years old and flat burned out from the grind. I was just going through the motions. I had been doing everything myself for so long that I didn't recognize I needed help to remain competitive as the sport evolved. I had no money and no retirement plan. I had given no consideration for the future due to being self-centered about what I wanted and not what was best for my family. I was now married and my wonderful, supportive wife wanted a kid. I had to start getting my ass in gear for our future and seek out a new career.

CHAPTER TEN

WHEN ONE DOOR CLOSES, ANOTHER OPENS

MY FRIEND JIM RAHM knew I was looking to change to a new career. Jim had been George Hurst's original partner in a two-car, dirt-floor garage in Ambler, Pennsylvania and had stayed in contact with George throughout the years. Though the Hurst Shifter Company that George founded was very successful, George was no longer a part of it at this point. George was working as a consultant for United Technologies Corporation (UTC), the 23rd largest company in the US at that time. UTC was a leading aerospace company that was getting into the automotive test equipment business for the OEMs and for the pending state and nationwide aftermarket emission testing programs. UTC hired Hurst to help them gain access to the OEMs and to better understand the automotive business. George noticed that while the company had brilliant engineers, they knew little about the automotive business. So he suggested that they hire some automotive people. He told the story to Jim Rahm, who then told me that I should check out UTC.

I came on board with UTC as the application development manager. I found myself working with aerospace engineers in developing the latest test equipment. I worked on computerized automotive test equipment, hot engine testing for OEMs' quality control, state and federal emission programs, and the nationwide aftermarket test equipment business. I also acted as a consultant

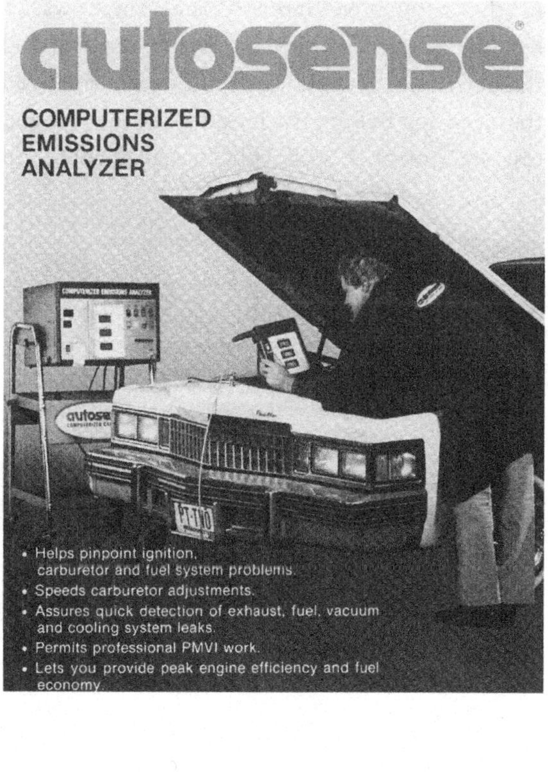

When he left racing behind at the end of 1973, Al took a job with United Technologies, where he developed programs to train mechanics in the use of modern diagnostic equipment. Al is featured on the company brochure.

for special diesel test equipment and the very first diesel electronic control system that they were developing as well. From the ground up I helped pioneer and develop the very first 4-gas exhaust analyzer for the automotive repair aftermarket.

I was already an automotive instructor and had tons of hands-on experience. But I knew I needed to extend my education in thermodynamics and physics while I was there, so I did. UTC brought in a professor from the Hartford Graduate Center to teach the group of aerospace engineers (me included) who were working on these automotive testers, the laws of thermodynamics, and energy conservation. I spent close to 15 years traveling all over the world, demonstrating equipment that I been instrumental in developing, while doing several state emission programs as well. I participated in hands-on public competitive demonstrations, diagnostic shoot-outs, and was a guest speaker to large and small groups on these subjects for years. After years of developing different new equipment, it became painfully apparent to me that what mattered was the man behind the tester, not the tester itself. It was just the beginning of electronic controls appearing in the automotive industry and there was a technology gap between the repair business and the newer electronic-controlled cars. I recognized the need for training in how to use the existing test equipment. I sold UTC's management on the idea of a program called CARES, an acronym that stood for Computerized Automotive Repair Educational System. It was an interactive automotive diagnostic training program, complete with a phone hotline service database for known repairs for specific makes and models with known symptoms. We had instructors who were resident experts in specific brands who you could consult when the database didn't help. When the result of a new diagnosis was achieved, it was then placed in the database for all participating members to pull from, kind of like a closed-loop feedback system.

It was impossible for anyone in the aftermarket to be an expert in all the makes and models of cars, and I always had empathy for mechanics, for obvious reasons. Even the new car dealers were only good at one brand, yet the aftermarket had to deal with all of them. I also became a member of the SAE (Society Automotive Engineers), which sets the automotive industry standards. I have been a member for over 35 years now and was a member

The CARES program was designed to bring mechanics up to speed on emission-controlled engines and the test equipment required to maintain them.

One of the training aids Al used in the CARES Program was this engine designed to allow students to view the combustion process as it took place.

of the SAE diagnostic subcommittee. Federal mandatory automotive emissions programs had ended Ford Motor Company's involvement in racing at one time, which greatly affected my racing career. In a twist of fate, here was a career opening for me doing these same mandatory emission tests. Thus, "one door closes, another opens." In the end, however, new management at UTC said that any division that was not aerospace, or was not doing a minimum of $500-plus million per year in sales would be sold off. Just like that it was the end of the automotive test division and me. So I was looking for another job, but this time I was smarter and had a ton of automotive experience both on and off the track.

The Phelon Years

I found a company near me called the R. E. Phelon Company, which made ignition flywheel assemblies for the lawn and garden industry. At the time Phelon was the largest ignition company in the world. They sold 17 million ignition modules and magneto flywheel assemblies per year. I went to them not for a job, per se, but to sell them on an idea I had for a true closed-loop ignition system. My closed-loop concept could overlap into the lawn and garden business, which was looking ahead at pending emission programs directed at them also.

The founder of the company had just died and his son Dale was now in charge. Dale wanted to start diversifying into other markets. My concept piqued his interest, so I got the title of director of new product development and marketing. Right away I was surprised and also concerned because my ignition concept was based on digital electronics, while Phelon's only expertise consisted of 40 years of analog electronics. Here I thought I was getting into a world-class, capable company, but they were in fact somewhat out of date in their understanding of digital electronics and computer controls. I was concerned that anything I wanted to do would be slowed, but Dale was willing to hire and/or have his engineers learn digital systems, so I thought everything might work out.

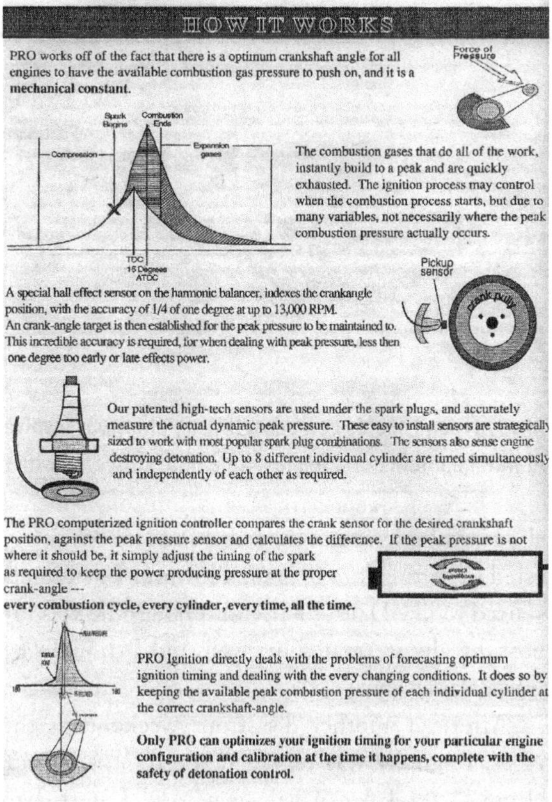

One of the ideas that Al brought to Dale Phelon's business was his closed-loop design. This would modernize the small-engine industry.

Along the way I noticed that Dale was interested in race cars. Apparently he always wanted to race cars, but both his dad and his former wife had frowned upon it. Now that his father had passed and left him the business, plus the fact that he also just gotten a divorce, he kind of "let the horse out of the barn," so to speak. Dale had a couple cars, but the one he was trying to race with was a 1969 Boss 302 Mustang in the SVRA vintage sports car Trans Am series that was growing in popularity. I watched out the corner of my eye as I saw some young guys working on the car in a big garage right next to the factory. I later learned they were having the engine done by a local speed shop/drag racer. Poor Dale couldn't even finish a race, let alone win. He kept breaking valves and valvetrain components, and he felt the horsepower of the engines was not that great either. He didn't know

Al soon found out his new boss, Dale Phelon, was involved in SCCA vintage racing with a Boss 302 Mustang. Al watched from the sidelines while Dale struggled with the car before finally stepping in to assist with the engine prep.

about my background in racing and I wasn't interested in getting into it again. However I genuinely started to like Dale as a friend. Though he was my boss, he always treated me well, and I thought he was a classy guy.

After yet another disastrous weekend at the races, I saw he was down in the dumps. I said, "Look, I think I can help you," and I started to tell him about my car racing experience. He lit up and asked me to jump in. He gave me carte blanche if I would help. I saw from his excitement that it would help our relationship if it all worked out, but I cautioned him that there are no guarantees in racing and it was hard work. The next day I asked the guys who were working on the car if there were any local dynos that we could use.

Dale's new shop in Aiken, SC soon evolved into a state-of-the-art facility for the preparation, maintenance, and storage of his race cars.

I then had the guys take the engine out of the car and put it on the dyno. It tested at 390 hp, which became our baseline. I then had the guys take the engine apart and lay out all the parts on the table to see what we had to work with. The two big items that jumped out to me were that the cam was way too big both in duration and lift. They had also used 1.8-ratio rocker arms—no wonder they were breaking valves. Also the typical boss 302 cylinder head ports were way too big, even though they had actually been partially filled.

I ordered a mid-range cam and some 1.6-ratio rocker arms from Crane. I learned from Sam's Cobra that you need mid-range torque as well as max horsepower for sports car racing. My goal was to keep it around 7,500 rpm for reliability, though 8,000-plus rpm seemed reasonable with this engine configuration. I spent a lot of time on the heads. I filled both the head ports and the trick Ford cross ram racing intake he had. I restricted the oil return channels in the head with adjustable jets. I believe I wound up with a .90" hole. This let the valve cover fill up with oil while running, cooling the springs yet still allowing oil to flow back. If the oil level got too high it returned through the pushrod holes.

"I left the rest of the engine the same, as it all looked good to go. I reassembled the engine and we took it to the dyno again. The dyno operator couldn't believe the horsepower number: 550-plus hp at 7,600 rpm. It was right on target. I was happy and lucky at the same time. These horsepower numbers were beyond what the dyno operator was used to seeing out of a 302. It was the highest-power 302 he had ever had on his dyno. I had the guys put the engine back in and they went to the races without me over the weekend. When Monday came there were some very happy faces in the garage, but that was nothing compared to when Dale came in. He was smiling from ear to ear and was a very happy man indeed, as he had won his first race. He went on to win the last race of the season as well.

Dale started to win right out of the gate the following year and dominated the class all season. I don't remember him losing.

Around this time frame Dale was having problems with both the unions and the EPA at his plant in Springfield, Massachusetts. He had five other plants and this one was becoming harder and harder for him to do business in, so he decided to move to and enlarge his other plant in Aiken, South Carolina. He offered me a moving package to follow him down and I took the deal.

He continued his winning ways with the Boss. He did most of his passing on the straightaways, though, as the car itself didn't handle as well as his competition. One day one of his competitors let him drive a Camaro that he wanted to sell, and Dale bought it on the spot. Dale felt it had much better 4-wheel weight balance and was much lighter overall. Even though it had less power it was just as fast in lap times. The Boss had more power, but carried too much weight over the front wheels with its all-iron engine and huge Cleveland heads. He did go back and forth between both cars and could turn comparable lap times with them, but he felt far more comfortable with the Camaro as it was easier for him to drive. In all fairness the Camaro was professionally built from the ground up and was lightened up as much as possible. The Boss was supposedly a George Follmer backup car that was never raced and was basically a stock Mustang.

Dale carried on his winning ways with the Camaro until we got to Sebring, where a terrible accident destroyed the Camaro and nearly took Dale's life. After the accident and during his recovery he decided that these old cars were not safe. The rules said you had to use original equipment, like the front suspension and brake components, and could not upgrade to safer, more reliable parts. As a result he wanted stop racing the old Trans Am cars and get into a safer car. At the time Roush Racing was absolutely dominating the SCCA Pro Racing Trans Am series and Dale expressed interest in trying to race one. I said hell, Jack Roush is an old friend. Dale was shocked when I told him that. Dale said, "You mean you could talk to him about getting one of his cars?" I said sure, I'll call him and I'll set it up.

We flew to Jack Roush's place and he gave us the grand tour of his businesses. Both Dale and I were impressed at the size and scope of his enterprises. He was into so many things at the same time; he maintained ties to all forms of motorsports. He even flew his own World War II-era P-51 warplane. In typical Roush fashion when he couldn't get parts like camshafts and valvetrain components, airframe and skin panels, and so on, he made them. He has become a real expert on Merlin engines. I bet I saw 25 engines in crates and in various state of assembly on behalf of customers all over the world. He turned a hobby into a big business.

Roush Racing had dominated the SCCA Trans Am pro series through the late 1980s to the mid 1990s, and that was why we were there. Dale purchased car #004, which was the only original-chassis car Roush had left.

Dale was in great shape, but he was around 55 years old and was racing against 21-year-old drivers. He had gone from the front of the pack to now trying to keep up and it was a humiliating experience for him. I had to hand it to Dale; he really wanted to improve and tried like hell, race after race. He was improving as a driver. He had a plant in Puerto Rico and because they had a Trans Am race there, he put the car and spare parts into containers and shipped it over. They asked for volunteers from Dale's plant to go to the race as helpers. It turned the whole event into a special one for Dale's employees. We all flew in and did our thing, not knowing what to expect. We were racing guys that we couldn't talk to because of the language barrier. The track was at an old airport and was very rough. The other drivers played rough too, as they didn't want some Gringo coming onto their turf and winning the race. Some were young and crazy drivers that Dale couldn't match, but he did come in 2nd. Bottom line, he lost the race, but they treated us very well. I'm not sure they would have done that if we had won, though. Dale was a good sport about it and we had another great experience.

After a winning season with his Boss 302 Mustang in SCCA vintage racing, Dale decided to step up to a custom-built Camaro racer for the series.

Dale loved racing so much he wanted to get better and hired a pro diver to show him the way. He needed to have the latest state-of-the-art cars if he wanted to win. Roush was now full-time into NASCAR and was not doing new SCCA cars any longer. Updating to a new car was the answer, and the only winning team building cars at that time was Rocket Sports, so Dale ordered two of their cars.

Even though we had a full-time crew of six for our Trans Am effort, it had become a grind and was no longer fun for me. I was just standing with a clipboard and mic most of the time during the race while he was having fun driving. Dale also bought professionally built engines for all the cars, so I didn't do that either.

I still enjoyed the more amateur, nostalgia-type racing that we started with. Pro racing was full-time hard work. We had been going at it for about 10 years and I wanted out. I asked Dale to find someone to replace me. He hired a

Traveling with the "wine and cheese racers," as he likes to call them, Al took advantage of Dale's private jet to travel to races. Gone were the days of driving a car hauler cross-country.

crew chief but after a short period Dale started to finally realize he was simply not going to be competitive in the pro racing series, mostly due to his age. However he still loved racing and went back to the SVRA nostalgia series with me. He just went back to his old ways by winning and having fun. He bought old Roush cars and raced them in SVRA.

SVRA also started a historical NASCAR series that really took off. There were plenty of NASCAR teams selling their old cars, while the "wine and cheese" racers loved them because they were relatively cheap, very safe, and parts were readily available. They were also born in America, and NASCAR was at its peak of popularity at the time.

Dale told me to go look at a bunch of NASCAR cars a guy had for sale so he could start racing one too. He wanted to sell Dale a Buick that had Chattanooga Chew livery. It was painted dark brown like tobacco and looked like shit to me. My wife Marge was with me that day and she liked (and so did I) a Wood Brothers #21 car the man had. The seller said he had a deposit on the car and didn't think it was available. I told him, if you want to sell the car right away, I have the trailer outside and we can close the deal right now. He called the guy who supposedly placed the deposit. He told the seller he didn't know when he would have the balance of the money. So he sold it to Dale and I took it home right there and then.

The car didn't have an engine and transmission, but it was a rolling chassis that had been built just

Al and Dale are all smiles in the pits at Sebring, where it appeared that they had a good chance for a win with the Camaro.

Those smiles turned to worried looks when Dale crashed the Camaro during the race, narrowly avoiding death.

for road racing. The first year of the nostalgia NASCAR rules limited the engines to 350 cubic inches, with period-correct iron block and heads in an attempt to control costs and speed. This was fun for me, as I put together another Boss 302 engine, only this time as a 350. One of the best times I ever had racing was when we went to the very first race of the new series in Sebring. I had talked him into going with just the pickup truck and car trailer that he had started with when he first went racing. Dale flew in while my wife and

After recovering from the crash at Sebring, Dale Phelon got right back on his horse and stepped up his game even more by purchasing a Roush Racing Mustang. Having a manufacturing plant in Puerto Rico was the perfect excuse for Dale, Al, and team to participate in a race there.

I drove the truck. Besides extra tires and wheels, we had no spare parts of any kind. If the car broke, we were going home. The car flew right out of the gate and was on the pole. There were a ton of GM and even Mopar cars there that day too. We cooked food on an open fire and just enjoyed ourselves. We kicked ass that day as Dale easily won the race. Dale was not the only one having fun that day—everyone was.

Easily one of the most interesting collector cars Dale had was the Roger Penske Can Am Lola that had been driven by world-famous Mark Donohue. It was the ultimate go-fast toy, a killer machine. It had a 500 cubic inch engine putting out around 800 hp with a weight of around 1,750 lbs., giving it a 2.2 lbs. per horsepower ratio. It was the wildest car I ever worked on. I learned firsthand what a truly amazing design it was, as we took it completely apart and put it back together again. All the structural integrity was in the rear. The engine and transaxle assembly was also the foundation for the rear suspension, which all tied to the roll bar. It had huge rear tires that were a full 20 inches across. The rest of the car was of a monocoque design. This meant there was no frame, per se, just .040" aluminum sheet metal formed into a structural configuration.

The monocoque was the reason the car was so light, but because there was no true frame, safety was thrown out the window. The driver's feet were exposed in any type of front-end accident. The fuel

Taking things to the next level, Dale purchased a state-of-the-art, professionally built Camaro and hired a pro driver as well. It wasn't long before both Dale and Al longed for the vintage-style racing again, though.

Getting back to vintage racing, Dale decided to indulge in the popular NASCAR vintage stock car races with this former Wood brothers, Neil Bonnet #21 Thunderbird. Al was delighted to be back in a more hands-on racing series.

tanks surrounded the driver as well. At that time these cars approached speeds of 250 mph. They had wings for downforce, and to counteract the natural tendency of the car's shape to fly off the ground. Dale was invited to race the car in a newly formed nostalgia Cam Am series that never went anywhere. Some of the other wine and cheese owners had them too, but when they started to race them, most needed diapers when they drove one in an actual race. You have to be careful what you wish for!

In the end Dale tried to become just a car owner. He jumped into the brand-new NASCAR truck series. He raced for two years but like so many other teams, he couldn't find a sponsor. He

Perhaps the most sophisticated race car that Dale ever owned was this Can-Am Lola.

Dale also owned a NASCAR truck series team for a time and scored one series win.

told me one day he looked in the mirror and said to himself, "I don't drive it, it costs me $125,000 per race for someone else to drive it, and if you win the prize money is only $40,000." So he decided to quit at the end of the year if things didn't change. Before the end of the year he did win one race, the Richmond 200. I was lucky enough to have been there with him that day and it was a genuine thrill to win one of these races. It was a huge achievement and it was especially nice to see him beat the Roush teams that day too. We both had come a long way since it all started.

Nevertheless Dale didn't find a sponsor that year. True to his word he quit the NASCAR truck series. He continued to participate in nostalgia racing against other teams of equal manpower and talent.

Things would change for me once again around this time period when I got a phone call from Jim Rahm regarding his latest project. He was building a very sophisticated, fiber composite, fully pressurized airplane. And, he wanted me to build him an engine for it ... holy shit!!!

CHAPTER 11

TAKING TO THE SKIES

WHEN JIM RAHM CALLED me about building an engine for his new high-tech airplane, he was referring to an automotive-type, water-cooled engine, not the standard air-cooled engine typically used in aircraft. His reasoning was that because he was building a next-generation airplane, he didn't want to install an engine design that was over 60 years old and had received only minimal updates. With my now-better knowledge of thermodynamics, I took his request very seriously. I realized that this was going to be a huge responsibility for me safety-wise. Reliability was far more important than horsepower or performance. After all, I didn't want to kill my best friend.

Jim had been interested in aviation since he was young. He and I went way back to the early 1960s, when he first started getting the experience it took to become the great pilot he would be. Some of our fun flying experiences were when I puked into his brand-new Stetson cowboy hat (he didn't want me to puke all over his new plane) when he took

Jim's Rahm's plane would be a cutting-edge, state-of-the-art design using all the best materials.

The aircraft would utilize an advanced Electronics Flight Instrument System from Archangel Avionics. The plane would have nearly all glass cockpit instrumentation with less reliance on conventional gauges.

me for a ride in his Beechcraft V-tail airplane, and the infamous 1963 Cobra story I told you earlier. I always got airsick when I was younger. I always loved the thrill and privilege of flight, but it was hard for me to get as excited about airplanes as much as Jim because of it.

This project was going to be a huge task. I had to take a leave of absence from Phelon. My original verbal agreement with Jim was that I was only going to do the engine. He was going to have it installed in the plane by a highly certified welder that he knew from the aviation community where he lived. Called Spruce Creek, it featured homes with hangers for private aircraft, and its own runway. The runway had started out as a WWII emergency landing strip. It later became a drag strip, and was the same place I raced and won with Louie Tomaselli in the 1958 Ford some 35 years earlier.

Jim's plane was high-tech, but it was a kit. It had to be built from the ground up. It took around three years to complete. It had to be inspected periodically by the FAA during the build process, complete with pictures and definitions of key safety areas. It was pressurized, so it had to be tested on the ground with 8 psi of pressure differential created by a huge air compressor. While looking for leaks during the pressurization test, it made all kinds of cracking noises as the pressure built up. The plane itself started to stretch, especially the pressurized door (it also had an inflatable door seal with its own air pump) as it started to actually bulge out. The 8 psi pressure was equal to tons of pressure across the entire hull of the plane. I never had seen this done before and it scared the hell out of me. I thought it was really a dangerous thing to do, because it might just blow up like a popping balloon.

Jim also had a beautiful, custom-made leather interior with all the trimmings done. The instruments were the latest glass cockpit design (minimal round gauges) as well.

Finally the entire body had to be sanded, filled, and fitted to perfection, before the show-quality paint job could be applied. This combined work took well over 3 months to complete. When finished, the plane was a piece of art.

While all of this was going on I started to look at what the aircraft engine builders were doing at the time. The most powerful standard aircraft engines at the time for a plane like this were 550 cubic inch, 6-cylinder engines with twin turbochargers, rated at 350 hp at 2,600 rpm, with cruise power of 262 hp at 2,300 rpm. They had to use bigger cubic inch engines to reach a certain power level, as it's difficult at best to make 350 hp at only 2,500 rpm. The crank was directly bolted to the propeller, and the propeller's peak efficiency was at a lower rpm relative to the engine, so designers were best served to create engines that made more power at these low rpms. Turboprop engines were another solution. These are just jet engines with a gear drive built in. Designers could engineer the optimum propeller and engine speed separately with no compromise, but they burn up to 40 percent more fuel at lower power levels, especially under 500hp, than a piston engine that can deliver the same horsepower. There was no free lunch!

It became apparent to me that the best solution was to use a gearbox or, in aviation terms, a propeller speed reduction unit (PSRU). This would permit a smaller engine to make more power at higher rpm, and also let the propeller run at an even lower, more efficient rpm at the same time. The big problem was we didn't have a suitable gearbox, and no one had any known good ones for sale that we knew of at the time.

The Engine

I wanted the engine to be safe, of high efficiency, and rated at a conservative power level. Water cooling was the only choice. There is nothing wrong with typical air-cooled engines; they have been in airplanes since aviation began. But they are just not as efficient as water-cooled engines. Because of the high certification cost of a new engine, dated air-cooled designs stick around. I'm sure that if the aviation market could justify the huge costs of developing and certifying a new engine, it would be water-cooled.

Jim was always a GM man and wanted to use a small block Chevy V-8 because they were universally known, small, relatively light, and made good power. They were also economical and readily available. I didn't see any problem using one, except there was no way we could make more power without raising the engine rpm. Like every project, we had a budget to stay within (even NASA has one), so cost was the main reason why we chose an existing water cooled automotive engine. I understood that automotive engines were designed for intermittent power duty cycles. They were never intended for the continuous power requirements of aircraft. To that end, this engine would have literarily no standard automotive components. I looked at it as an aircraft engine built on an automotive platform. I planned on using nothing but heavy-duty, robust components (not lightweight racing parts) with extra cooling capabilities in key areas. These included sophisticated electronic controls that could be calibrated for the continuous power that an aircraft application requires, bringing along safety and reliability, too.

Because this aircraft was experimental, we were unencumbered by certification requirements faced by aircraft engine companies. We followed the philosophy of, "we retain the right to be smarter as we gain knowledge and experience, and also reserve the right to take advantage of the latest technology as it becomes available." As a result, we would have different versions of both the engine and the installation packaging. I set the engine power targets at a conservative 420hp takeoff

Al utilized heavy duty, aftermarket internal engine components that would be familiar to any hotrodder. Additional modifications were performed to enhance lubrication and thus longevity.

During testing the Siamese middle exhaust port design of the 350 Chevrolet cylinder heads caused oil temperatures to reach such a level that "coking" (burning) of the oil occurred.

Cylinder heads with re-designed exhaust ports were cast to correct the overheating problem, and steel rings were fashioned to replace conventional gaskets on the exhaust side of the cylinder heads.

This size comparison between a conventional turbocharger on the right and one required to allow the engine modified for aircraft use to produce the goal of 420 horsepower at take off and 315 horsepower at cruising altitude on the left. The increased size of the turbo demanded additional heat and space considerations.

power with a 315 hp cruise. Controlling piston speed is crucial for long-term reliability, so I set a goal not to exceed 2,500 feet per minute at cruise (around 3,600 rpm). The maximum known engine BMEP load (brake mean effective pressure) would be around 200 psi, which was at 6 psi boost pressure at around 4,650 rpm. The turbo was not used to boost power per se, (though I did steal 6 psi for maximum takeoff power for a short time only, just like the aircraft engines do), but to offset atmospheric pressure loss at high attitudes.

Because I didn't have a magneto, I had to have totally redundant electronic controls, with two batteries, two computers, two electric fuel pumps, two crank pickups and two sets of engine sensors, all for safety. Air-cooled engines uses dual-plug heads as a way to deal with detonation under high-temperature ambient conditions and heavy load. Dual plugs helped eliminate end-gas detonation phenomenon by speeding up the combustion time. With a water-cooled engine, I only used one spark plug per cylinder, with one direct ignition coil per plug, and I eliminated the troublesome magneto. With computer controls I had all kinds of calibration options for both the fuel and ignition that the ancient magnetos and mechanical fuel injection systems could only dream about. It was kind of neat to calibrate the engine on the dyno and in the air with the computer. I also had data recording so I could look at engine information anytime Jim took a flight.

I originally had just agreed to work on his engine, but the whole thing would soon become a nightmare for me. The original guy who was

Al's engine design is shown here on the dynamometer during testing. Mounting and cooling the engine in the airframe would create unanticipated problems for Al.

Mounting the automotive-based V-8 engine and all its components into the nose of the aircraft posed many unique challenges. Shown at the front of the engine is the gearbox, the component that caused the most difficulty in the project.

supposed put the engine in gave up trying to figure it all out, and I got stuck with it. There would be many versions of the final engine installation. One part always seemed to interfere with the other due to the incredibly small space and heat considerations that required separation and insulation of components. Just the assorted water, air, and oil heat exchanges and directional air baffling was a task unto itself. Of course you had to be able to service everything as well. The uncovered exhausts were impossibly close to the radiators and intercoolers. If you wrapped the exhaust with thermal blankets you ultimately overheated the pipe and welds and they failed,

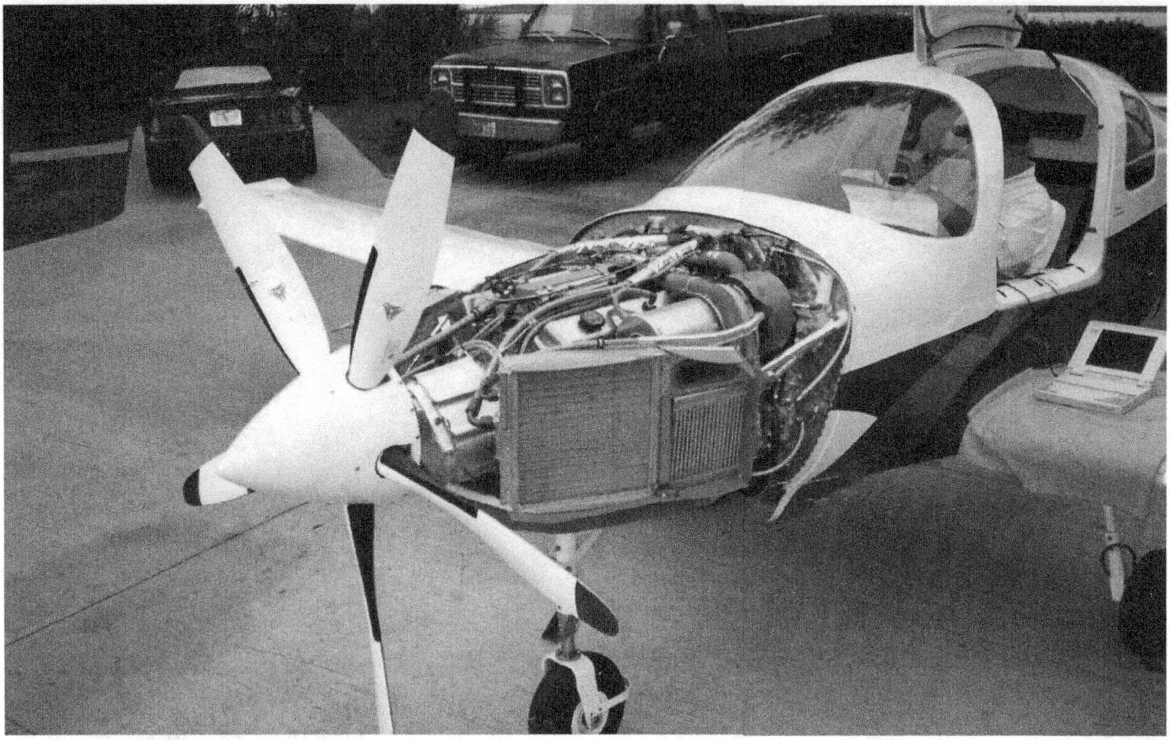

This photo shows the engine finally in place. Additional shielding of exhausts and modifications to the cowling for cooling would be required before the plane was completed and flown.

especially after many cycles of expansion and contraction. Instead I used a titanium heat shield that stopped the line-of-sight radiation problem, but that still let the air cool the pipes and associated components through convection cooling.

Aside from the assorted packaging concerns, I also had to make sure that the center line of the gearbox shaft came out on the center line of the cowl to maintain the aerodynamics of the plane's original design. The center line issue created another problem; I had to lower the engine into the motor mount to the point where the oil pan would be so shallow that there was not enough reserve oil in it. In turn, I then had to make yet another custom part, an aluminum oil pan with side blisters on it to add oil capacity. The added capacity was essential to let the oil stand and de-aerate before it was recycled. The aeration issue came from the engine itself as well as the turbo and gearbox. It would whip the oil into foam and if you didn't let it sit for a short time, the oil could become completely foamed. The whole process was like putting 5 lbs in a 2 lb bag.

The Gearbox

The gearbox became the nightmare of the entire project. It was not only a technological challenge, but we would learn the hard way that pioneers get arrows in their backs. Jim found an aviation guy (most aviation guys are passionate and want to be a part of an interesting project somehow) who owned a machine business. He said he had a great and innovative design for a gearbox. Jim cut the deal with a minimal input from me. I was just happy we could get someone to take the job on. However, about two years and lots of money later, we were still waiting for the promised gearbox. When we finally received it the plane was close to being done and so was the engine. As we modeled the installation with a mock up gearbox they provided, the so-called innovation didn't work. We did manage some high-speed taxi work with it but that was all.

It proved almost impossible to source a reliable gearbox to adapt the automotive-based engine to an airplane. The fixture shown in this photo was used to test the gearboxes before use. The inability to procure a sufficient number of gearboxes would eventually doom the project.

The technology challenge was the resonant frequency of the power pulses from the engine versus the propeller's constant-speed flywheel effect. The resonant frequency was created by the engine producing 1 power pulse every 90 degrees and the flywheel effect of the approximately 100-lb propeller plus the resistance of the blades biting the air. The propeller absorbed all the horsepower we could produce and acted as a heavy but constant-rpm load opposing the engine's pulses. As the horsepower increased at lower rpms, the peak amplitudes of these very high-power pulses started bending the gears and shafts.

You could either make the gears and shafts extremely robust to tolerate these peak loads, thus making them heavier than required, or you could dampen the peaks and reduce the loads. As the rpms increased, the power frequency did smooth out and the resonant frequency issues were no longer a problem. The damping part of the first gearbox's design didn't work. We wound up in court and never got any satisfaction, thus wasting years and tons of money. We then found another guy who actually had a gearbox for sale, but it was the only one he had. He too was an aviation enthusiast, and a good engineer with a good design. He also had the capability to design and make a gearbox, at least. We bought the gearbox at a premium and started to fly with it. The problem was we needed more than one. I felt that since he had a design that worked, how hard could it be to get more? So we ordered more ... but just like with the first guy, we had to wait close to two years and never received a gearbox. It turned out he was a good designer and prototype guy, but he was a one-man band and production was not his thing. He started to order castings of the housing and gears etc. but there was always something that stopped us from getting finished gearboxes.

The First Flight

We had to do a lot of ground testing before flying and boy, did it take a lot of time! We were in Florida in the summer and it was very hot. This created a challenge in keeping the engine cool with the cowl on. Each air inlet opening was only about 6 inches long. The air had to go around a lot of corners to come though the radiators and finally through the bottom cowl outlets. When standing still there was a lot of restriction for any free-flowing air.

When we first started to test, the new instrumentation failed as well, mostly because of the Florida heat. We had a test pilot doing the first flight work as he had flown many other similar planes. He was contracted for a few days and then was scheduled to leave, so everyone was getting inpatient and frustrated as he tried to take off, with no idea of how hot the engine really was. I was startled when he firewalled the throttle and went zooming by. As he approached the end of the runway the engine started to misfire at the same time the airplane left the ground. My heart was in my mouth as this once-fun project all of a sudden became very serious.

The engine continued to misfire as the plane slowly lifted up to treetop level and circled around the runway to immediately land again. By the time he landed the badly overheated engine had taken

The unique, high-tech, V-8 powered airplane was featured on the cover of *Sport Aviation* magazine.

out the top ring lands of two pistons. We were damn lucky, as it could have been much worse. After we fixed up the engine again, we started doing higher-speed runs and everything started to work OK. But there was still an overheating issue when standing still or doing low-speed taxi work. As a result I added electric fans for ground cooling, which helped a lot. However, it didn't fix the problem completely. I was forced to add additional gille vent opening directly next to the turbocharger, which was a major source of heat. This opening also permitted a huge amount of heat to exit after the engine shut down. The combination of these solutions finally gave us what we were looking for.

Jim Rahm was also presented with the August Raspet Award for outstanding contributions to the advancement of light airplane design.

EAA '97 OSHKOSH July 30-August 5 1997

EAA Aviation Center
P.O. Box 3086
Oshkosh, WI
54903-3086
414-426-4800
FAX 414-426-4828

May 23, 1997

Mr. James D. "Jim" Rahm
209 Cessna Blvd., #2
Daytona Beach, FL 32124-6860

Dear Jim:

Each year EAA presents the August Raspet Award to the individual(s) that makes outstanding contributions to the advancement of lightplane design. As you can see by the attached list of Award winners, it is a highly recognized group of EAAers and individuals who have become the leaders in the aircraft design and homebuilding movement. This year I am pleased to announce that you and Al Joniec have been selected as the 1997 recipients of the August Raspet Award.

Both of you are deserving of this award through your efforts to take the personal airplane to a higher technological level. You have proven once again that the freedom to innovate allowed by the homebuilt regulations is a benefit to society.

This award will be presented at the Theater in the Woods (during the Annual EAA Convention) on Thursday evening, July 31st, at approximately 7:55 p.m. I would appreciate it if you could confirm your acceptance with my secretary Shirley McQuillan (414-426-6585). I hope that you and Al can be with us. Please be back stage by 7:30 p.m.

I look forward to hearing from your. Congratulations!

With best regards,

EAA OSHKOSH '97

Tom Poberezny
Convention Chairman

Enclosure

McQ

Gateway to Aviation
45th annual EAA Fly-In Convention

THE TECHNICAL CHALLENGE

The biggest misconception that the average guy has (including me) is that an engine is an engine, and therefore can apply to any vehicle—WRONG! I still clearly remember when I was young and first started to work on cars and wanted to understand more about them. I asked a grizzled old truck mechanic if he could tell me why the engines were so big in trucks but made only 250 hp compared to a much smaller car engine that made 300 hp. He said he didn't know. But he did know that if you had two trucks, one with a car engine, placed a big load in both trucks and then drove over a mountain range, only the one with the truck engine would come down the other side. I never forgot that.

It took me many years before I would start to understand about that the term "continuous power" really meant the ability to produce max rated horsepower all the time. Car engines were rated for intermittent max power. Continuous power requirements apply to aviation, heavy-duty trucks, and stationary power and marine applications. Racers, especially drag racers, had the opposite power requirements. They have sprint and/or intermittent max power requirements with short-term life expectancy before overhaul.

All engines, including external combustion steam engines, which the internal combustion engine evolved from, are heat engines. If you want to make more power, you have to make more heat. When you make more heat than you can control in the critical hottest areas, like in the exhaust valve bridge of the head, failure ensues through detonation. And I don't mean the water temperature, either. The water temp is just the average temperature of the coolant and doesn't indicate the real surface temperature on the components. There is a speed limit to how much heat can pass through any material into the cooling system.

That is why drag racers can get away with murder. Heat over time is the issue. It takes a huge amount of heat energy over time for a 600 lb engine, for example, to reach temperature equilibrium of all of its critical limits. Even aircraft engines have a time limit for their rated max power usage and are typically only rated for continuance power at cruise power settings. The real problem in aviation is that if you want to go fast, you have to go high. I believe you can go around 23 percent faster for every 1,000 feet of altitude due to reduced aero drag of the thinner air with the same power. The problem is you *have* to keep going faster because the thinner air won't hold the plane up. I believe we had to go at least 300 mph or so because the plane would stall at peak altitude. However, the turbo starts to make a huge amount of heat when compressing the air and tends to make the engine prone to detonate. For every 20-degree increase to the inlet air temperature, you needed 1 additional point of octane rating in your fuel. At an altitude of 25,000 feet, the ambient air pressure is is only 5.3 psi, which means the turbo has to make a pressure ratio of about 3 to 1. At 33,000 feet the atmospheric pressure is 3.4 psi, and this requires close to a 5 to 1 pressure ratio, comparable to a Top Fuel dragster. When you get to these kinds of pressure ratios, the inlet temps start to go off the map. They can go over 300 degrees even when you are seeing -56 degrees air temperature at altitude at the turbo inlet, thus the need for high-efficiency intercoolers. Remember, it's all about heat management; making any more heat than you can control over time makes problems for the engine very quickly.

Engine Test and Prototype

I finally settled on my core engine configuration and started to identify and buy parts. It would be an all-aluminum engine with a high-strength 4340 crankshaft. The crank was not lightened for more strength and for more rotating mass to dampen out the power pulses. The rods were 4130 standard weight but I had them drilled through using EDM (electro deposition machined) to pass oil up to the wrist pin and squirt the underside of the center of the piston to guarantee they were getting cooled.

I started out with heavier hypereutectic pistons because they had up to 15 percent ceramic in the

At the Oshkosh aviation show, the plane was given a place of honor next to the supersonic Concorde airliner.

aluminum alloy. This made them harder and longer-wearing, especially in the top ring land area. But I updated to standard forged pistons after we had a failure when we encountered overheating problems. The center of the piston is the hottest part and needs the most oil cooling. By having the rods drilled for oil I cooled the hottest part of the pistons the right way, directly through the connecting rods. This also cooled the top of the rod, lubricated the wrist pins, and sprayed oil on the pistons. I call this the rifle approach, versus the shotgun approach, which uses oil squirters from the block.

One of the biggest weaknesses of the small block Chevy engines was that the two center exhaust ports are siamesed, which causes a major hot spot. Standard heads have no water cooling between the ports, so we got some spread-port heads made to our specs that had water coolant passages in them to help deal with the problem. The heads also had smaller ports for the lower rpm we worked at. I also modified the block water passage to direct as much water to this center area as practicable. However, even after taking these steps, it still got so hot from continual power that we experienced oil coking in the valve cover area. I finally stopped cleaning it as the coking itself acted as a heat shield for further coking.

This unique high-heat situation was because of the turbo temps of 1,500-plus degrees and high back pressure of up to 23 psi. Unlike a normally aspirated engine, the exhaust valve and port never had a cooling cycle. The higher in attitude you go, the higher the temperatures and back pressure at the same power. At 2,500 feet even the inlet engine temps got up to 300 degrees because of the pressure ratio the turbo had to maintain. To get these unacceptable inlet temps down we had to have

Here are eight engines being prepared for customers of Engine Air.

The company also became involved in twin-engine aircraft as well. The first photo shows an engine installation and the second shows Al posing proudly with the finished product.

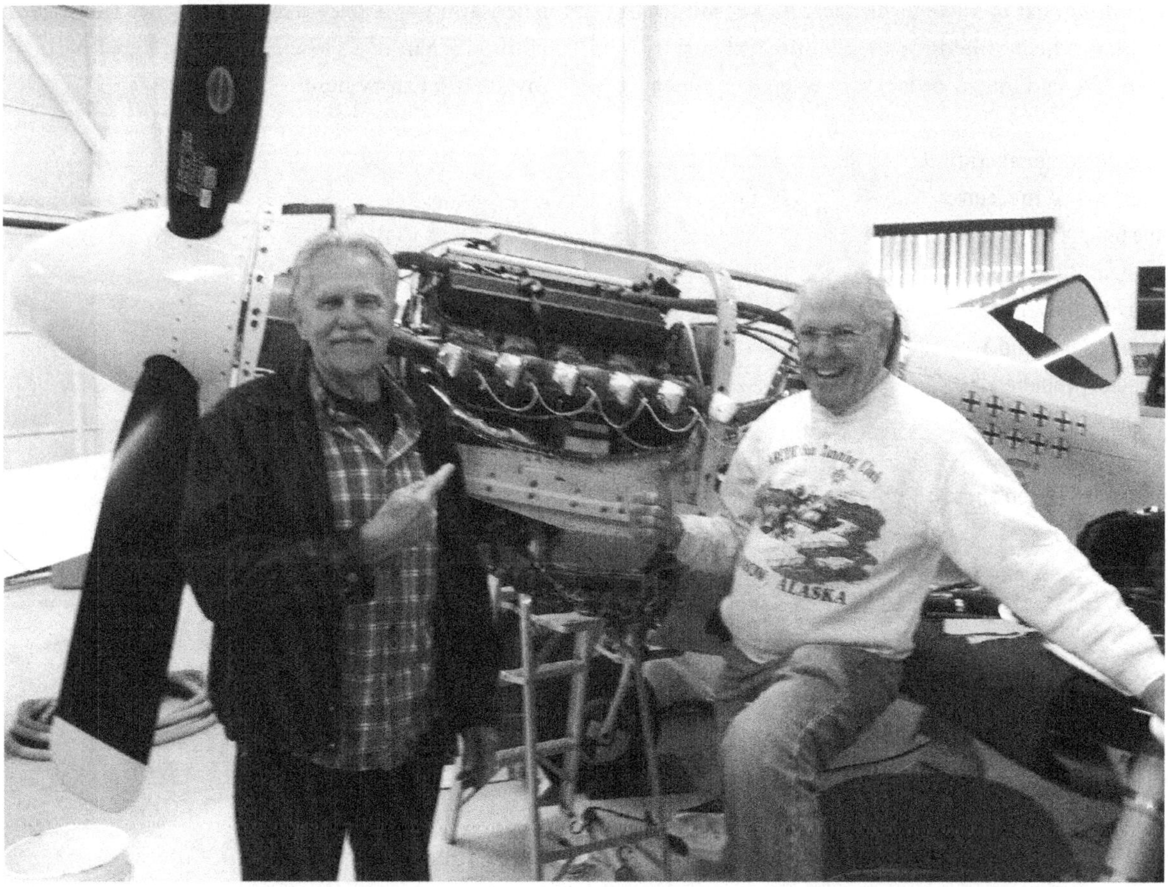

The naturally aspirated engines are used to power seventy-percent size replicas of the iconic P-51 fighter planes of World War II and Korea.

very efficient intercoolers. The combined total heat load that the engine cooling system had to reject was way beyond reasonable. The water cooling system also cooled the oil with a water-to-oil heat exchanger. However, a major source of heat that I didn't figure on initially was that the oil also cooled the gearbox (about 2 percent power loss was typical in a gearbox) which equated to about 20,000 BTUs an hour added to the engine cooling requirements. I later added a small air-to-oil heat exchanger for the engine oil before going into the gearbox.

The extremely high, continuous radiant heat coming from the exhaust system next to the closely mounted radiator and other components was a nightmare. After 750 degrees, radiant heat exceeded convection heat transfer and was a problem child. No matter how much air you passed over it, the radiation was dominant, kind of like standing next to a fire or furnace. As a result I had to make heat shields for the entire exhaust system. We had lots of orders for our engine package, but after 2-plus years of waiting for the gearbox, some deposits had to be refunded while others continued to wait. I did get a few gearboxes finished and got a few customers flying. But after a while we started to lose all of them and I will end my aviation story here.

Modified P-51s are still widely used in airplane racing, and one such aircraft, powered by an Al Joniec/Robert Abernathy engine, won a Bronze Award in the Sports Class at the prestigious Reno Air Races.

This customer aircraft was the last to receive the unique V-8 engine design. Thanks to the use of lightweight components and a special propeller, the plane set an unofficial speed record for propeller-driven aircraft of 432 MPH.

EPILOGUE

I'M NOT SURE HOW, but I fell in love with cars when I was young and, while I didn't know it at the time, it most likely saved me. This passion evolved into racing, which led me to a career and a lifetime love affair with engines. I also now realize that I was lucky to be on the ground floor of the so-called muscle car era as well. It turned out that racing would became one of the best times in my life. I quickly found out that winning really helps to keep it fun; losing, not so much. I genuinely loved it and when you're young and full of piss and vinegar you can handle the required workload. Even when I finally got sponsored by Ford, I had to step it up even further to stay competitive with ongoing research and development in the world of "speed secrets." This was life-changing, as I also had to learn about resource management, public relations, and the logistics of getting a professionally prepared car to weekly events widely spaced across the county. Eventually you pay the price as the pace starts to burn you out.

At the time I didn't know the true definition of technology; I just worked my ass off, that's all. I constantly tried everything and made as many runs down the track as I could. I let the time slips show me the way. Though I had tons of hands-on experience, it was years later that I learned how engines really worked, and understood the laws of physics, thermodynamics and the laws of energy conservation. Racing aside, I have been a member for the Society of Automotive Engineers (where all the true automotive technology comes from) for close to 40 years now. I can objectively look back on it all in amazement, as I lived through some of the same "hits and misses" of 60-plus years' worth of evolution that has brought us today's sophisticated cars and engines.

It is with both sadness and awe that I view the next technological evolution taking place. Though they will never become totally obsolete, the internal combustion engine in automobiles will slowly be replaced with electric motors. There will also be totally autonomous, computer-controlled cars. In fact there is currently a driverless racing series.

My continual quest to understand and learn about the internal combustion engine continues to this day. I recently tried to do a couple of versions of an engine myself. The first one was a 1-stroke engine, meaning that every time the piston went up or down there was a power stroke. I was looking for the most power density possible. I took an existing single-cylinder engine and modified it. Because I am a fan of the external combustion steam engine, I used the double-acting piston concept and applied it to an internal combustion engine. I had a custom piston, a cylinder with all of the associated porting, and cross-head rod assembly made. Next I got a couple of orbital fuel injectors that inject air as well as fuel into both the top and bottom combustion chambers. I adapted an aftermarket fuel and ignition computer control system and calibrated it.

I got the engine running and even though I didn't test it on a dyno, it felt and sounded powerful to me. My naivete in attempting this type of engine quickly surfaced as the enemy of the high power-density engine is heat, and/or the ability to control it. Boy, did I learn about heat when I melted the aluminum piston. I replaced it with one made out of steel, but it quickly got so hot that it also failed. I also noticed the cylinder walls were also getting scuffed from the high heat. I had a tiger by the tail.

I had water cooling of the cylinder and could control that, but the piston was getting hot every 180 degrees and never got a cooling cycle like a 2-cycle engine does. I was forced to start researching how I could cope with these overheating issues. Lo and behold, 1-stroke engines had been attempted before, the classic, "if you don't read history you're doomed to repeat it" scenario. What an education I got. Apparently these earlier innovators also had the same problems. I did do a drawing for a dynamic liquid cooling system for the piston, but I ran out of time and money and shelved the project.

I was still hot to do an advanced engine project. I had a potential investor who liked my 1-stroke concept. I confessed to him that I didn't do my homework on the engine before I started on it. I said I had not concluded if the concept would ultimately work, but I had a potential solution.

While doing research I found that the best minds in the industry felt there was still a lot of room for improvement to the internal combustion engine. The Holy Grail was "constant volume during combustion," and if it could be achieved. All engines have the problem of the piston outrunning the combustion process. This requires additional fuel to maintain combustion and meaningful cylinder pressure. By having the combustion completed while the piston is at or near top-dead center (TDC), less fuel energy is required to make the same pressure as the displacement is not increasing before combustion is completed. All the energy can be expanded to do work prior to the exhaust valve opening.

Armed with this very credible engineering information, my focus now shifted, as did the investor's. I spent considerable time thinking of the ways and means of how to achieve it. I went back to the investor with my concept and he accepted it. We started a company called NGEN. I again used an existing engine and retrofitted it with my mechanical innovation concept. My concept was to mechanically pause the piston during approximately 40 degrees of crank rotation. While the piston remained at TDC and allowed combustion to be completed, the crank continued to rotate. After several failures I did in fact achieve constant volume during combustion. I instrumented the engine with a combustion chamber sensor to measure pressure during combustion. This pressure trace normally shows a peak profile, but with my concept the pressure profile is flat at and around TDC. I felt I now had enough credibility to take it to another level. I solicited several high-end investors to no avail. I even went to the second-largest engine development company, FEV, and got enthusiastic feedback. However, it would take multiple millions of dollars to see it to the end. Quite frankly I didn't understand this game. I soon found out that literally dozens of concept engines are out there, all promising to save the world, and all out there competing for the same investor money. I was constantly told, show me your dyno data results to support your claims and we will talk. Because I had already exhausted all of my time and money on the project to get it to where it was, I was forced to quit. I was getting smarter, but that wouldn't pay the mortgage. However I did get a patent in my name just in case things changed.

In an attempt to stay active I have begun to do engine seminars, as I still enjoy teaching what I know to anyone who is willing to listen. I still build some aircraft engines with Robert Abernathy, Jr. of Abby's Performance engines, with whom I have been working for more than 20 years on many of my engine projects.

ABOUT THE AUTHOR

BY AL JONIEC

ALONG THE ROAD OF LIFE (in this case a quarter-mile at a time), you meet certain people that you feel exceptionally comfortable with, and Charlie Morris is one of the few I feel lucky to have met. Initially what attracted me to him was our common passion for Ford race cars, especially the so-called factory muscle cars of the '60s—an era that, in my opinion, will never be repeated.

I call the 1960s the age of discovery, as you just didn't know what you were going to see at the track. There were literally no rules. You have to remember unlike today, there was no internet and/or social media. Hell, there was hardly any printed media for a while, and when it did come out it was cherished. It could be like telling stories around the campfire of olden days. Word of mouth was all there was sometimes. Every week there was another trick car and/or innovation that evolved to today's extremely fast Funny Cars. This evolution had to come from somewhere, and in my opinion today's car enthusiast is somewhat desensitized to this fact. Most take for granted

This is one of my favorite pictures of Charlie and me. I gave him an original picture and the actual front bat emblem of the *Batcar* at a Ford car show. Because Charlie had given so much to the sport, I felt I would be remiss if I hadn't at least given him this small token of my appreciation.

the blood, sweat, and tears, high cost, and general pain of the evolutionary process.

I, for one, consider myself a student of the internal combustion engine. I have learned, through the process of studying the past, that we are all collectively standing on the shoulders of giants who have come before us. I admire Charlie for attempting to record this exciting era, for those who are interested, and the people who made it happen.

He told me way back that his interest was so great that he would like to write a book about it all (see Charlie's letter and note the incredible detail of his handwriting). This made me admire him even more, as little did I dream he would actually follow through with it. This is an example of the kind of follow-through and stick-to-it-tiveness he has to actually write a book. It's hard at best to write a single book, and now he has authored several of them. Charlie is not doing it to become rich; he's doing it to preserve the history based only on his passion. Charlie is not only a great car guy, author, and automotive historian, but a patriot who you would be lucky to be stuck in a foxhole with, should the fecal matter ever hit the fan.

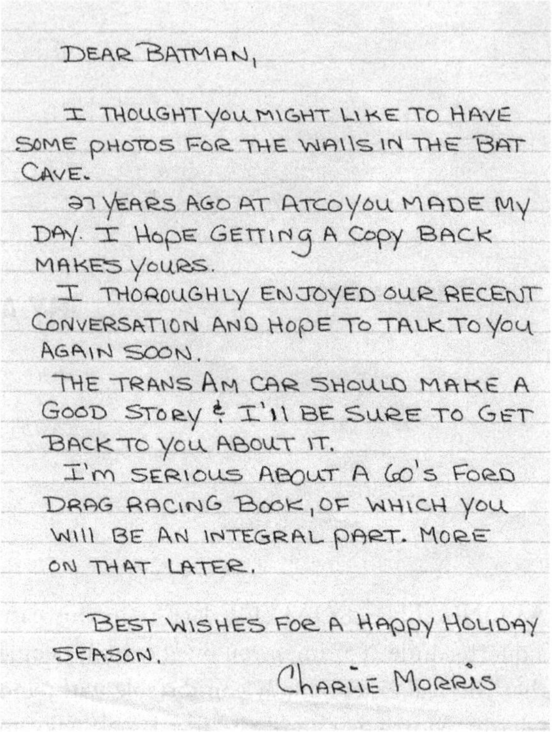
DEAR BATMAN,

I THOUGHT YOU MIGHT LIKE TO HAVE SOME PHOTOS FOR THE WALLS IN THE BAT CAVE.

27 YEARS AGO AT ATCO YOU MADE MY DAY. I HOPE GETTING A COPY BACK MAKES YOURS.

I THOROUGHLY ENJOYED OUR RECENT CONVERSATION AND HOPE TO TALK TO YOU AGAIN SOON.

THE TRANS AM CAR SHOULD MAKE A GOOD STORY & I'LL BE SURE TO GET BACK TO YOU ABOUT IT.

I'M SERIOUS ABOUT A 60'S FORD DRAG RACING BOOK, OF WHICH YOU WILL BE AN INTEGRAL PART. MORE ON THAT LATER.

BEST WISHES FOR A HAPPY HOLIDAY SEASON.

CHARLIE MORRIS

If I had to describe Charlie Morris the man in one word, it would be **integrity**. His prior occupation in law enforcement has given him a unique blend of the real-world experience and understanding of the most complex system of all: people. These experiences have grounded him with the innate ability to recognize, cope with, and tolerate morality issues better than most. As a result, the pure innocence of animals has also made Charlie an animal rights activist—a friend of those that have no friends. His work in this area is to me far more important than car books. The animal rights crusade that he takes on makes me humble to my core, and I admire him for it more than he knows.

Now happily married to his wife Gale, Charlie is not only a polite, all-around good guy, but has no ego and is a pleasure to be around and work with. I feel honored that he has chosen to write this book about me, but what I feel really great about is calling him a good friend.

Made in the USA
Monee, IL
22 September 2026

64075042R00096